Structural Welding Code— Steel

Third Edition

Superseding
AWS D1.1-75

and

AWS D1.1-Rev 1-76
AWS D1.1-Rev 2-77

Prepared by
AWS Structural Welding Committee

Under the Direction of
AWS Technical Activities Committee

Approved by
AWS Board of Directors

Effective January 1, 1979

AMERICAN WELDING SOCIETY, INC.
2501 N.W. 7th Street, Miami, Fla. 33125

Library of Congress Number: 78-64959
International Standard Book Number: 0-87171-163-X

American Welding Society, 2501 N.W. 7th Street, Miami, FL 33125

Note: By publication of this code the American Welding Society does not insure anyone utilizing the code against liability arising from the use of such code. A publication of a code by the American Welding Society does not carry with it any right to make, use, or sell any patented items. Each prospective user should make an independent investigation.

Printed in the United States of America

Contents

CONTENTS

Personnel

F. S. Adams	American Institute of Steel Construction
T. Agic	Kaiser Steel Corporation
W. G. Alexander	New York State Department of Transportation
C. A. Baker	Kansas City Structural Steel Company
E. M. Beck	Law Engineering Testing Company
M. H. Bell	Consultant
J. T. Biskup	Canadian Welding Bureau
O. W. Blodgett	The Lincoln Electric Company
J. A. Bradley	J. A. Bradley & Associates
R. W. Christie	Hardesty and Hanover
L. Colarossi	Pittsburgh-Des Moines Steel Company
D. E. Conklin	Nuclear Power Products, Inc.
M. F. Couch	Bethlehem Steel Corporation
H. F. Crick	Brown and Root, Inc.
M. V. Davis	American Welding Society
T. J. Dawson	Naval Facilities Engineering Command, Department of the Navy
T. J. Downey	Sverdrup & Parcel
T. G. Ferrell	Belmas-Jovel, Inc.
A. R. Fronduti	Gamble's Inc., A Trinity Company
G. A. Gix	American Bridge Division, U.S. Steel Corporation
M. F. Godfrey	Federal Highway Administration, DOT
R. H. Goldsmith	Ammann & Whitney
R. R. Graham, Jr.	U.S. Steel Corporation
R. J. Harris*	Consultant
C. E. Hartbower	Federal Highway Administration, DOT
T. R. Hensley	Flint Steel Corporation
G. J. Hill	Michigan Department of State Highways and Transportation
E. Holby	Fluor Pioneer, Inc.
A. L. Johnson	American Iron and Steel Institute
A. J. Julicher	A. J. Julicher & Associates
H. A. Krentz	Canadian Institute of Steel Construction
R. A. LaPointe	Stone and Webster Engineering Corporation
G. C. Lee	J. Ray McDermott & Company
P. W. Marshall	Shell Oil Company
P. E. Masters	Consultant
T. McCabe	Inryco, Incorporated
W. McGuire	Cornell University
W. A. Milek, Jr.	American Institute of Steel Construction
W. C. Minton	Southwestern Research Institute
W. H. Munse	University of Illinois
E. F. Nordlin	California Department of Transportation
C. W. Ott	U.S. Steel Corporation
A. E. Pearson	Tucker Steel, Inc.

*Resigned March 1978

C. A. Pestotnik	Iowa Department of Transportation
C. W. Pinkham	S. B. Barnes & Associates
W. R. Pressler	Pittsburgh Testing Laboratory
F. H. Ray	Ohio Department of Transportation
C. R. Rea	Texas Department of Highways and Transportation
F. A. Reickert	Hazelet & Erdal
D. E. H. Reynolds	Dominion Bridge Company Ltd.
P. F. Rice	Concrete Reinforcing Steel Institute
J. E. Roth	Schneider Sheet Metal, Inc.
J. P. Shedd	Howard, Needles, Tammen and Bergendoff
G. A. Shenefelt	American Bridge Division, U.S. Steel Corporation
J. L. Simmons	AMCA International
D. L. Sprow	J. Ray McDermott & Company
J. R. Stitt	J. R. Stitt & Associates
L. Tall	Lehigh University
J. D. Theisen	Brown and Root Western Hemisphere Marine
W. E. West	St. Joseph Structural Steel Company
J. C. Williams	Combustion Engineering, Inc.

AWS Structural Welding Committee and Subcommittees

Main Committee

J.T. Biskup, Chairman
F. H. Ray, Vice-Chairman
O.W. Blodgett, Vice-Chairman
M.V. Davis, Secretary
T. Agic
W. G. Alexander
J. A. Bradley
R.W. Christie
L. R. Colarossi

M. F. Couch
H. F. Crick
T. J. Dawson
T. J. Downey
T. G. Ferrell
G. A. Gix
R. H. Goldsmith
R. R. Graham, Jr.
C. E. Hartbower

T. R. Hensley
R. J. Harris**
A. J. Julicher
R. A. LaPointe
P. W. Marshall
P. E. Masters*
W. A. Milek, Jr.
W. H. Munse*
E. F. Nordlin*

A. E. Pearson
C. A. Pestotnik
W. R. Pressler
C. R. Rea
F. A. Reickert
D. Reynolds
D. L. Sprow
J. R. Stitt
L. Tall*

Subcommittee 1 on Design

R.W. Christie, Chairman
W. A. Milek, Vice-Chairman
R. H. Goldsmith
A. L. Johnson
A. J. Julicher
H. A. Krentz
W. McGuire
P. W. Marshall
W. H. Munse

E. F. Nordlin
F. H. Ray
C. R. Rea
F. A. Reickert
J. P. Shedd
J. L. Simmons
L. Tall
W. E. West

Subcommittee 2 on Prequalified Procedures

L. R. Colarossi, Chairman
T. G. Ferrell, Vice-Chairman
T. Agic
D. H. Conklin

T. J. Dawson
A. R. Fronduti
R. H. Goldsmith
D. E. H. Reynolds

Subcommittee 4 on Workmanship

A. E. Pearson, Chairman
W. R. Pressler, Vice-Chairman
T. Agic
W. G. Alexander
C. A. Baker
M. F. Couch
T. J. Downey
A. R. Fronduti
G. A. Gix
M. F. Godfrey

P. E. Masters
W. A. Milek
C. A. Pestotnik
F. H. Ray
C. R. Rea
D. E. H. Reynolds
J. P. Shedd
J. R. Stitt
J. C. Williams

Subcommittee 3 on Qualification

D. L. Sprow, Chairman
E. Holby, Vice-Chairman
J. A. Bradley
D. H. Conklin
M. F. Couch
T. G. Ferrell
R. R. Graham

C. E. Hartbower
P. E. Masters
C. A. Pestotnik
J. R. Stitt
J. D. Theisen
W. E. West

Subcommittee 5 on Inspection

W. G. Alexander, Chairman
G. J. Hill, Vice-Chairman
F. S. Adams
J. A. Bradley
L. R. Colarossi
H. F. Crick
T. J. Downey
E. Holby

C. E. Hartbower
A. J. Julicher
W. C. Minton
E. F. Nordlin
A. E. Pearson
W. R. Pressler
G. A. Shenefelt

Subcommittee 6 on Tubular Structures

H. F. Crick, Chairman
P. W. Marshall, Vice-Chairman
E. M. Beck
O.W. Blodgett
R. R. Graham

G. C. Lee
T. J. Dawson
D. L. Sprow
L. Tall
J. D. Theisen

Subcommittee 8 on Sheet Steel

O. W. Blodgett, Chairman
A. L. Johnson, Vice-Chairman
T. McCabe
W. McGuire

C. W. Pinkham
J. E. Roth
J. L. Simmons
J. C. Williams

Subcommittee 7 on Reinforcing Bars

R. A. LaPointe, Chairman
M. F. Godfrey

C. W. Ott
P. Rice

*Advisory Member
**Resigned March 1978

Foreword

In 1928 the first edition of the Code for Fusion Welding and Gas Cutting in Building Construction was published by the American Welding Society. Since then, nine other editions of the Code have been published. During the latter part of July 1934, a committee was appointed to prepare specifications for the design, construction, alteration, and repair of highway and railway bridges by fusion welding. The first edition of the specifications was published in 1936, followed by seven other editions.

Until 1963 there were two committees, one for the Building Code and one for the Bridge Specifications. These two major committees recognized the community of interest in establishing a better level of structural welding standardization in the industry and had been cooperating for some time. In June 1963, these two committees were abolished and the Structural Welding Committee was organized. This committee is concerned with the preparation of standards and the promulgation of sound practices for the application of welding to the design and construction of structures. Since its organization, the committee has prepared the Building Code and the Bridge Specifications.

The 1972 edition was prepared to cover structural welding in general, along with specific requirements for buildings, bridges, and tubular structures. This consolidation eliminated the duplication in previous editions by combining the Code and the Specifications into a single document. The 1975 edition published revisions, errata, and editorial changes. The present edition of the Code includes additions and changes necessary to keep it current with the practices of and the requirements for structural welding.

Sections 1 through 7 constitute a body of rules for the regulation of welding in steel structures. Sections 8, 9, and 10 contain additional rules applicable to specific types of structures—buildings, bridges, and tubular structures—and should be used as a supplement to the first seven sections. For general structural welding of statically loaded structures where no specific code or specification is applicable, Section 8 is recommended.

Certain shielded metal arc, submerged arc, gas metal arc, and flux cored arc welding procedures plus certain types of joints have been thoroughly tested and have a long record of satisfactory performance. These procedures and joints are designated as prequalified, may be employed without presentation of further evidence (1.3), and include most of those that are commonly used. However, the purpose of defining prequalified procedures and joints is not to prevent the use of other procedures as they are developed. When other processes, procedures, or joints are proposed, they are subject to the applicable provisions of Sections 2, 3, and 4 and shall be qualified by tests in accordance with the requirements of Section 5. In the same section, the requirements for the qualification of welders, welding operators, and tackers are also given.

This code does not concern itself with such design considerations as the arrangement of parts, loading, and the computation of stresses for proportioning the load-carrying members of a structure and their connection. Such considerations, it is assumed, are covered elsewhere in a general code or specification such as a Building Code, AISC Specification for the Design, Fabrication and Erection of Structural Steel for Buildings, American Association of State Highway and Transportation Officials, Standard Specifications for Highway Bridges, American Railway Engineering Association Specifications for Steel Railway Bridges, or other specifications prescribed by the owner.

Fatigue testing has demonstrated that any abrupt discontinuity of section and stress path is a factor adversely affecting the strength of members subject to cyclic loading. Gradual rather than sudden transitions of sections should be employed, and for the same reason, butt welds are preferable to fillet welds.

In the case of old structures, material of questionable weldability may have been used (including wrought iron or high-strength structural silicon or nickel steels). Accordingly, it is advisable when making repairs to an old structure to obtain samples of the material and to make laboratory tests for developing the proper welding procedure and weld values.

Comments or inquiries pertaining to this code are welcome. They should be addressed to: Secretary, Structural Welding Committee, American Welding Society, 2501 N.W. 7th Street, Miami, Florida 33125.

Preface

The 1979 edition of the Structural Welding Code represents a departure from previous editions in an effort to resolve the continuing problem of promptly providing the user with current revisions to the Code. Henceforth, the Code is to be issued annually in a single bound volume. This should provide for orderly revision to the Code without the confusion of the voluminous colored pages of revisions issued previously. The growing need of the Structural Welding Committee to consider revisions in keeping with expanding technology means that in the forseeable future there should be considerable and substantial changes to the Code. The Committee is continually striving to keep pace with the advances in construction technology as well as the rules of the many state and federal regulatory bodies.

The user should be aware that there were no 1978 revisions issued, nor was there a 1978 edition of the Code. This 1979 Code supersedes the 1977 revisions to the 1975 Code (AWS D1.1-REV2-77). Designating this edition as the 1979 Code makes the designation more representative of the issue date.

The 1979 edition contains new provisions, revisions of current material, and rearrangements of material from the 1977 revisions. Some of the more important aspects of these changes are outlined in the following paragraphs.

Changes in Code Requirements. Where changes have been made in Code requirements, a double vertical line appears in the margin immediately adjacent to the paragraph affected. A major editorial change in the material will be indicated by a single vertical line immediately adjacent to the paragraph affected. This continues the style and format that has previously been used in the revisions of the 1975 Code. Changes to tables and figures as well as new tables or new figures have not been so indicated.

Effective Date. For the first time, the 1979 Code will include an effective date, January 1, 1979. This is the date when all users can reasonably be expected to have the Code in hand and to have reviewed its provisions. The Structural Welding Committee has added the effective date at the request of many of the users of the Code. Since it is expected that the Code will be issued annually in mid-fall, future editions will have an effective date of January 1st of the year following the issue date.

New Drawings. All of the illustrations have been redrawn for this edition of the Code. For the first time, the prequalified joint details contain all of the dimensional information, including assembly tolerances. There has been a consolidation of the 104 sketches of prequalified joint details shown in the previous edition to the 40 sketches that appear in this issue. The user should find this arrangement helpful, since all necessary information relating to dimensional aspects of the joint detail appear in one place. All drawings have been edited to exhibit clarity and conciseness. The weld areas in all the Code sketches have been shaded to define the weld area, which the user should find helpful.

New Tables. All of the tables have been reviewed and where necessary, revised and all tables have been reset for clarity and usefulness. In Section 5, Qualification, new tables have been added to display all procedure qualification provisions to tabular form. These tables should be helpful to the user in the application of Code provision to procedure qualification.

In Appendix E, two tables have been added which compile the provisions necessary to establish prequalified joint welding procedures, and list the Code provisions that may be modified when a joint welding procedure is established by test. A check list has been provided to assist the user in preparing prequalified joint welding procedures. This check list includes all Code provisions that must be observed when preparing prequalified joint welding procedures.

Rearrangement of Provisions. To make the Code more coherent, the workmanship-type provisions that were previously found in Section 4, Technique, i.e., caulking, arc strikes, weld cleaning, groove weld termination, and groove weld backing have been moved to Section 3, Workmanship. Stress relief heat treatment has been moved from Section 3, Workmanship, to Section 4, Technique.

Index. Much time and effort have been expended to increase the usefulness of the index, which contains more entries than were previously listed. In this edition of the Code the entries are referred to by paragraph number rather than by page number. This change should enable the users to locate the item of interest in minimum time.

Structural Welding Code—Steel

1. General Provisions

1.1 Application

1.1.1 This code covers welding requirements applicable to any type of welded structure. It is to be used in conjunction with any complementary code or specification for the design and construction of steel structures. It is not intended to apply to pressure vessels or pressure piping. Requirements that are essentially common to all structures are covered in Sections 1 through 7, while provisions applying exclusively to buildings (static loading), bridges (dynamic loading), or tubular structures are included in Sections 8, 9, and 10, respectively.

1.1.2 All references to the need for approval shall be interpreted to mean approval by the Building Commissioner,[1] the Engineer,[2] or the duly designated person acting for and in behalf of the owner on all matters within the scope of this code. Hereinafter, the term Engineer will be used, and it is to be construed to mean the Building Commissioner, the Engineer, or the duly designated person who acts for and in behalf of the owner on all matters within the scope of this code.

1.2 Base Metal

1.2.1 Approved Base Metals. The base metals to be welded under this code are carbon and low-alloy steel commonly used in the fabrication of steel structures. Steels complying with the specifications listed in 8.2, 9.2, and 10.2, together with special requirements applicable individually to each type of structure, are approved for use with this code. Steels other than those listed in 8.2, 9.2, and 10.2 may be used provided the provisions of 8.2.3, 9.2.4, or 10.2.3 are complied with.

1.2.2 Thickness Limitations. The provisions of this code are not intended to apply to welding base metals less than 1/8 in. (3 mm) thick. Where base metals thinner than 1/8 in. are to be welded the requirements of AWS D1.3, Specification for Welding Sheet Steel in Structures, should apply. When used in conjunction with AWS D1.3, the applicable provisions of this code shall be observed.

1.3 Welding Processes

1.3.1 Shielded metal arc welding (SMAW), submerged arc welding (SAW), gas metal arc welding (GMAW) (except short circuiting transfer), and flux cored arc welding (FCAW) procedures which conform to the provisions of Sections 2, 3, and 4, in addition to Sections 8, 9, or 10, as applicable, shall be deemed as prequalified and are therefore approved for use without performing procedure qualification tests.

1.3.2 Electroslag (ESW) and electrogas[3] welding may be used provided the procedures conform to the applicable provisions of Sections 2, 3, and 4 and the contractor qualifies them in accordance with the requirements of 5.2.

1.3.3 Stud welding may be used provided the procedures conform to the applicable provisions of 4.21 through 4.27.

1.3.4 Other welding processes may be used provided they are qualified by applicable tests as prescribed in 5.2, and approved by the Engineer. In conjunction with the tests, the joint welding procedures and limitation of essential variables applicable to the specific welding process must be established by the contractor developing the procedure. The range of essential variables shall be based on documented evidence of experience with the process, or a series of tests shall be conducted to establish the limit of essential variables. Any change in essential variables outside the range so established shall require requalification.

---

1. The term "Building Commissioner' refers to the official or bureau, by whatever term locally designated, who is delegated to enforce the local building law or specifications or other construction regulations.

2. The Engineer is the duly designated person who acts for and in behalf of the owner on all matters within the scope of this code.

3. The term "electrogas welding" as used in this code refers to either gas metal arc welding-electrogas (GMAW-EG) or flux cored arc welding-electrogas (FCAW-EG), or to both.

1.4 Definitions

The welding terms used in this code shall be interpreted in accordance with the definitions given in the latest edition of AWS A3.0, Terms and Definitions, supplemented by Appendix I of this code.

1.5 Welding Symbols

Welding symbols shall be those shown in the latest edition of AWS A2.4, Symbols for Welding and Nondestructive Testing. Special conditions shall be fully explained by added notes or details.

1.6 Safety Precautions

Safety precautions shall conform to the latest edition of ANSI Z49.1, Safety in Welding and Cutting, published by the American Welding Society.

1.7 Standard Units of Measurement

The values stated in U.S. customary units are to be regarded as the standard. The metric (SI) equivalents of U.S. customary units given in this code may be approximate.

2. Design of Welded Connections

Part A
General Requirements

2.1 Drawings[4]

2.1.1 Full and complete information regarding location, type, size, and extent of all welds shall be clearly shown on the drawings. The drawings shall clearly distinguish between shop and field welds.

2.1.2 Drawings of those joints or groups of joints in which it is especially important that the welding sequence and technique be carefully controlled to minimize shrinkage stresses and distortion shall be so noted.

2.1.3 Contract design drawings shall specify the effective weld length and, for partial penetration groove welds, the required effective throat, as defined in 2.3 and 10.8. Shop or working drawings shall specify the groove depths (S) applicable for the effective throat (E) required for the welding process and position of welding to be used.

2.1.3.1 It is recommended that contract design drawings show complete joint penetration or partial joint penetration groove weld requirements. The welding symbol without dimensions designates a complete joint penetration weld, as follows:

complete joint
penetration weld (CP)

The welding symbol with dimensions above or below the arrow designates a partial joint penetration weld, as follows:

(E_1) partial joint
(E_2) penetration weld

where

E_1 = effective throat, other side
E_2 = effective throat, arrow side

4. The term "drawings" refers to plans, design and detail drawings, and erection plans.

2.1.3.2 Special groove details shall be specified where required.

2.1.4 Detail drawings shall clearly indicate by welding symbols or sketches the details of groove welded joints and the preparation of material required to make them. Both width and thickness of steel backing shall be detailed.

2.1.5 Any special inspection requirements shall be noted on the drawings or in the specifications.

2.2 Basic Unit Stresses

Basic unit stresses for base metals and for effective areas of weld metal for application to buildings, bridges, and tubular structures shall be as shown in Part B of Sections 8, 9, and 10, respectively.

2.3 Effective Weld Areas, Lengths, and Throats

2.3.1 Groove Welds. The effective area shall be the effective weld length multiplied by the effective throat.

2.3.1.1 The effective weld length for any groove weld, square or skewed, shall be the width of the part joined, perpendicular to the direction of stress.

2.3.1.2 The effective throat of a complete joint penetration groove weld shall be the thickness of the thinner part joined. No increase is permitted for weld reinforcement.

2.3.1.3 The effective throat of a partial joint penetration groove weld shall be the depth of chamfer, less 1/8 in. (3.2 mm) for grooves having an included angle less than 60 deg, but not less than 45 deg at the root of the groove, when deposited by shielded metal arc or submerged arc welding, or when deposited in the vertical or overhead welding positions by gas metal arc or flux cored arc welding.

3

The effective throat of a partial joint penetration groove weld shall be the depth of chamfer, without reduction, for grooves

(1) Having an included angle of 60 deg or greater at the root of the groove when deposited by any of the following welding processes: shielded metal arc, submerged arc, gas metal arc, flux cored arc, or electrogas welding; or

(2) Having an included angle not less than 45 deg at the root of the groove when deposited in flat or horizontal positions by gas metal arc or flux cored arc welding.

2.3.1.4 The effective throat thickness for flare groove welds when filled flush to the surface of the solid section of the bar shall be as shown in Table 2.3.1.4.

(1) Random sections of production welds for each welding procedure, or such test sections as may be required by the Engineer, shall be used to verify that the effective throat is consistently obtained.

(2) For a given set of procedural conditions, if the contractor has demonstrated that he can consistently provide larger effective throats than those shown in Table 2.3.1.4, the contractor may establish such larger effective throats by qualification.

(3) Qualification required by (2) shall consist of sectioning the radiused member, normal to its axis, at midlength and terminal ends of the weld. Such sectioning shall be made on a number of combinations of material sizes representative of the range used by the contractor in construction or as required by the Engineer.

2.3.1.5 The minimum effective throat of a partial penetration groove weld shall be as specified in Table 2.10.3.

2.3.2 Fillet Welds. The effective area shall be the effective weld length multiplied by the effective throat thickness. Stress in a fillet weld shall be considered as applied to this effective area, for any direction of applied load.

2.3.2.1 The effective length of a fillet weld shall be the overall length of the full-size fillet, including end returns. No reduction in effective length shall be made for either the start or crater of the weld if the weld is full size throughout its length.

2.3.2.2 The effective length of a curved fillet weld shall be measured along the center line of the effective throat. If the weld area of a fillet weld in a hole or slot computed from this length is greater than the area found from 2.3.3, then this latter area shall be used as the effective area of the fillet weld.

2.3.2.3 The minimum effective length of a fillet weld shall be at least four times the nominal size, or the size of the weld shall be considered not to exceed one fourth its effective length.

2.3.2.4 The effective throat shall be the shortest distance from the root to the face of the diagrammatic weld. See Appendix A. Note: See Appendix B for formula governing the calculation of effective throats for fillet welds in skewed T-joints. A convenient tabulation of measured

legs (w) and acceptable gaps (g) related to effective throats (t) has been provided for dihedral angles between 60 deg and 135 deg.

2.3.3 Plug and Slot Welds. The effective area shall be the nominal area of the hole or slot in the plane of the faying surface.

2.3.4 The effective throat of a combination partial joint penetration groove weld and a fillet weld shall be the shortest distance from the root to the face of the diagrammatic weld minus 1/8 in. (3.2 mm) for any groove detail requiring such deduction (see Appendix A).

Part B
Structural Details

2.4 Fillers

2.4.1 Fillers may be used in

2.4.1.1 Splicing parts of different thicknesses.

2.4.1.2 Connections that, due to existing geometric alignment, must accommodate offsets to permit simple framing.

2.4.2 A filler less than 1/4 in. (6.4 mm) thick shall not be used to transfer stress but shall be kept flush with the welded edges of the stress-carrying part. The sizes of welds along such edges shall be increased over the required sizes by an amount equal to the thickness of the filler (see Fig. 2.4.2).

2.4.3 Any filler 1/4 in. (6.4 mm) or more in thickness shall extend beyond the edges of the splice plate or connection material. It shall be welded to the part on which it is fitted, and the joint shall be of sufficient strength to transmit the splice plate or connection material stress applied at the surface of the filler as an eccentric load. The welds joining the splice plate or connection material to the filler shall be sufficient to transmit the splice plate or connection material stress and shall be long enough to avoid overstressing the filler along the toe of the weld (see Fig. 2.4.3).

2.5 Partial Joint Penetration Groove Welds

Partial joint penetration groove welds subject to tension normal to their longitudinal axis shall not be used where design criteria indicate cyclic loading could produce fatigue failure. Joints containing such welds, made from one side only, shall be restrained to prevent rotation.

Table 2.3.1.4
Effective throats of flare groove welds

Flare-bevel-groove welds	Flare-V-groove welds
All diam bars	
5/16 R	1/2 R*

Note: R = radius of bar.
*Except 3/8 R for GMAW (except short-circuiting transfer) process with bar sizes 1 in. (25.4 mm) diam and over.

Part C
Details of Welded Joints

2.6 Joint Qualification

2.6.1 Joints meeting the following requirements are designated as prequalified:

(1) Conformance with the details specified in 2.7 through 2.10 and 10.13.

(2) Use of one of the following welding processes in accordance with the requirements of Sections 3, 4, and 10 as applicable: shielded metal arc, submerged arc, gas metal arc (except short circuiting transfer), or flux cored arc welding.

2.6.1.1 Joints meeting these requirements may be used without performing the joint welding procedure qualification tests prescribed in 5.2.

2.6.1.2 The joint welding procedure for all joints welded by short circuiting transfer gas metal arc welding (see Appendix D) shall be qualified by tests prescribed in 5.2.

2.6.2 Joint details may depart from the details prescribed in 2.9 and 2.10 and in 10.13 only if the contractor submits to the Engineer his proposed joints and joint welding procedures and at his own expense demonstrates their adequacy in accordance with the requirements of 5.2 of this code and their conformance with applicable provisions of Sections 3 and 4.

2.7 Details of Fillet Welds

2.7.1 The details of fillet welds made by shielded metal arc, submerged arc, gas metal arc, or flux cored arc welding to be used without joint welding procedure qualifications are listed in 2.7.1.1 through 2.7.1.5 and detailed in Figs. 2.7.1 and 10.13.1.5.

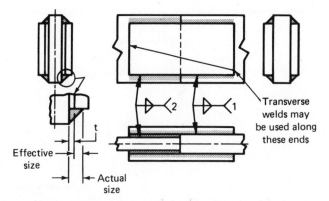

Note: The effective area of weld 2 shall equal that of weld 1, but its size shall be its effective size plus the thickness of the filler t.

Fig. 2.4.2—Fillers less than 1/4 in. thick

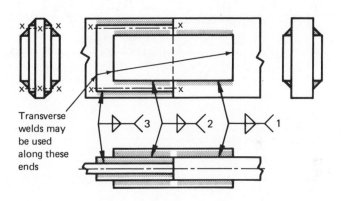

Notes:

1. The effective area of weld 2 shall equal that of weld 1. The length of weld 2 shall be sufficient to avoid overstressing the filler in shear along planes x-x.

2. The effective area of weld 3 shall at least equal that of weld 1 and there shall be no overstress of the ends of weld 3 resulting from the eccentricity of the forces acting on the filler.

Fig. 2.4.3—Fillers 1/4 in. or thicker

2.7.1.1 The minimum fillet weld size, except for fillet welds used to reinforce groove welds, shall be as shown in Table 2.7.

2.7.1.2 The maximum fillet weld size permitted along edges of material shall be

(1) The thickness of the base metal, for metal less than 1/4 in. (6.4 mm) thick (see Fig. 2.7.1, detail A).

(2) 1/16 in. (1.6 mm) less than the thickness of base metal, for metal 1/4 in. (6.4 mm) or more in thickness (see Fig. 2.7.1, detail B), unless the weld is designated on the drawing to be built out to obtain full throat thickness.

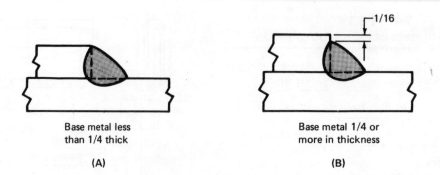

Base metal less
than 1/4 thick

(A)

Base metal 1/4 or
more in thickness

(B)

Maximum size of fillet weld along edges

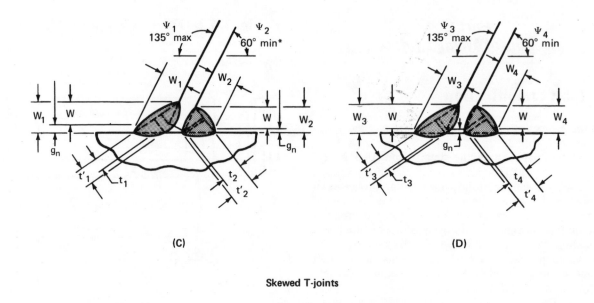

(C)

(D)

Skewed T-joints

Note: $t_{(n)}$, $t'_{(n)}$ = effective throats dependent on magnitude of gap (g_n). See 3.3.1. Subscript (n) represents 1, 2, 3, or 4.

*Angles smaller than 60 degrees are permitted; however, in such cases, the weld is considered to be a partial joint penetration groove weld.

Fig. 2.7.1—Details for prequalified fillet welds

2.7.1.3 Fillet welds in holes or slots in lap joints may be used to transfer shear or to prevent buckling or separation of lapped parts. These fillet welds may overlap, subject to the provisions of 2.3.2.2. Fillet welds in holes or slots are not to be considered as plug or slot welds.

2.7.1.4 Fillet welds may be used in skewed T-joints having a dihedral angle (ψ) of not less than 60 deg nor more than 135 deg (see Fig. 2.7.1, details C and D).

2.7.1.5 The minimum length of an intermittent fillet weld shall be 1-1/2 in. (38.1 mm).

2.7.1.6 Minimum spacing and dimensions of holes or slots when fillet welding is used shall conform to the requirements of 2.8.

2.8 Details of Plug and Slot Welds

2.8.1 The details of plug and slot welds made by the shielded metal arc, gas metal arc (except short circuiting transfer), or flux cored arc welding processes are listed in 2.8.2 through 2.8.8 and may be used without performing the joint welding procedure qualification prescribed in 5.2, provided the technique provisions of 4.28 or 4.29, as applicable, are complied with.

2.8.2 The minimum diameter of the hole for a plug weld shall be no less than the thickness of the part containing it plus 5/16 in. (8.0 mm), preferably rounded to the next greater odd 1/16 in. (1.6 mm). The maximum diameter of the hole for a plug weld shall not be greater than 2-1/4 times the depth of filling.

2.8.3 The minimum center-to-center spacing of plug welds shall be four times the diameter of the hole.

2.8.4 The length of the slot for a slot weld shall not exceed ten times the thickness of the part containing it. The width of the slot shall be no less than the thickness of the part containing it plus 5/16 in. (8.0 mm), preferably rounded to the next greater odd 1/16 in. (1.6 mm), nor shall it be greater than 2-1/4 times the depth of filling.

2.8.5 Plug and slot welds are not permitted in quenched and tempered steels.

2.8.6 The ends of the slot shall be semicircular or shall have the corners rounded to a radius not less than the thickness of the part containing it, except those ends which extend to the edge of the part.

2.8.7 The minimum spacing of lines of slot welds in a direction transverse to their length shall be four times the width of the slot. The minimum center-to-center spacing in a longitudinal direction on any line shall be two times the length of the slot.

2.8.8 The depth of filling of plug or slot welds in metal 5/8 in. (15.9 mm) thick or less shall be equal to the thickness of the material. In metal over 5/8 in. thick, it shall be at least one-half the thickness of the material but no less than 5/8 in.

Table 2.7
Minimum fillet weld size for prequalified joints

Base metal thickness of thicker part joined (T)		Minimum size of fillet weld*		
in.	mm	in.	mm	
T≤1/4	T≤ 6.4	1/8**	3	Single pass welds must be used
1/4<T≤1/2	6.4<T≤12.7	3/16	5	
1/2<T≤3/4	12.7<T≤19.0	1/4	6	
3/4<T	19.0<T	5/16	8	

*Except that the weld size need not exceed the thickness of the thinner part joined. For this exception, particular care should be taken to provide sufficient preheat to ensure weld soundness.

**Minimum size for bridge applications is 3/16 in.

Legend for Figs. 2.9.1 through 2.10.1

Symbols for joint types
B—butt joint
C—corner joint
T—T-joint
BC—butt or corner joint
TC—T- or corner joint
BTC—butt, T-, or corner joint

Symbols for base metal thickness and penetration
L—limited thickness—complete joint penetration
U—unlimited thickness—complete joint penetration
P—partial joint penetration

Symbols for weld types
1—square-groove
2—single-V-groove
3—double-V-groove
4—single-bevel-groove
5—double-bevel-groove
6—single-U-groove
7—double-U-groove
8—single-J-groove
9—double-J-groove

Symbols for welding processes if not shielded metal arc
S—submerged arc welding
G—gas metal arc welding
F—flux cored arc welding

2.9 Complete Joint Penetration Groove Welds

2.9.1 Complete joint penetration groove welds made by shielded metal arc, submerged arc, gas metal arc (except short-circuiting transfer), or flux cored arc welding in butt, corner, and T-joints which may be used without performing the joint welding procedure qualification test prescribed in 5.2 are detailed in Fig. 2.9.1 and are subject to the limitations specified in 2.9.2.

2.9.1.1 All complete joint penetration groove welds made by short circuiting transfer gas metal arc welding (see Appendix D) shall be qualified by the welding procedure qualification tests prescribed in 5.2.

2.9.2 Dimensional Tolerances. Dimensions of groove welds specified on design or detailed drawings may vary from the dimensions shown in Fig. 2.9.1 only within the following limits.

2.9.2.1 The specified thickness of base metal or weld effective throat is the maximum nominal thickness that may be used.

2.9.2.2 The groove angle is minimum; it may be detailed to exceed the dimensions shown by no more than 10 degrees.

2.9.2.3 The radius of J-grooves and U-grooves is minimum. It may be detailed to exceed the dimensions shown by no more than 1/8 in. (3 mm). U-grooves may be prepared before or after fit-up.

2.9.2.4 Double-groove welds may have grooves of unequal depth, but the depth of the shallower groove shall be no less than 1/4 of the thickness of the thinner part joined.

2.9.2.5 The root face of the joint shall be as dimensioned in Fig. 2.9.1 with the following variations permitted:

(1) For SMAW, GMAW, or FCAW it may be detailed to exceed the specified dimension by no more than 1/16 in. (2 mm). It may not be detailed less than the specified dimension.

(2) For submerged arc welding the specified root face of the joint is maximum.

2.9.2.6 The root opening of the joints is minimum. It may be detailed to exceed the specified dimension by no more than 1/16 in. (2 mm), except that the root opening of closed joints for submerged arc welding shall be detailed as zero (no variation).

2.9.3 Groove preparations detailed for prequalified shielded metal arc welded joints may be used for prequalified gas metal arc or flux cored arc welding.

2.9.4 Corner Joints. For corner joints the outside groove preparation may be in either or both members, provided the basic groove configuration is not changed and adequate edge distance is maintained to support the welding operations without excessive melting.

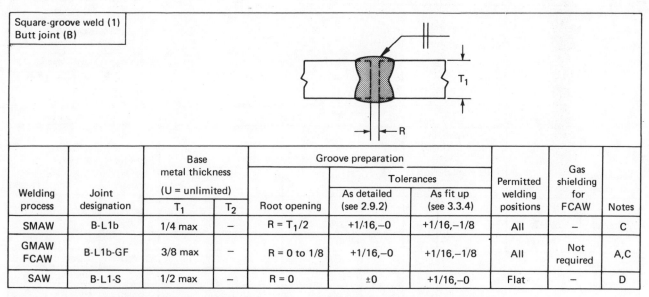

Square-groove weld (1)
Butt joint (B)
Corner joint (C)

| Welding process | Joint designation | Base metal thickness (U = unlimited) | | Groove preparation | | | Permitted welding positions | Gas shielding for FCAW | Notes |
		T₁	T₂	Root opening	Tolerances As detailed (see 2.9.2)	As fit up (see 3.3.4)			
SMAW	B-L1a	1/4 max	–	R=T₁	+1/16,–0	+1/4,–1/16	All	–	–
	C-L1a	1/4 max	U	R=T₁	+1/16,–0	+1/4,–1/16	All	–	–
GMAW FCAW	B-L1a-GF	3/8 max	–	R=T₁	+1/16,–0	+1/4,–1/16	All	Not required	A

Square-groove weld (1)
Butt joint (B)

| Welding process | Joint designation | Base metal thickness (U = unlimited) | | Groove preparation | | | Permitted welding positions | Gas shielding for FCAW | Notes |
		T₁	T₂	Root opening	As detailed (see 2.9.2)	As fit up (see 3.3.4)			
SMAW	B-L1b	1/4 max	–	R = T₁/2	+1/16,–0	+1/16,–1/8	All	–	C
GMAW FCAW	B-L1b-GF	3/8 max	–	R = 0 to 1/8	+1/16,–0	+1/16,–1/8	All	Not required	A,C
SAW	B-L1-S	1/2 max	–	R = 0	±0	+1/16,–0	Flat	–	D

Note A: Not prequalified for gas metal arc welding using short circuiting transfer. Refer to Appendix D.
Note C: Gouge root before welding other side.
Note D: Welds must be centered on joint.

Fig. 2.9.1—Prequalified complete joint penetration groove welded joints

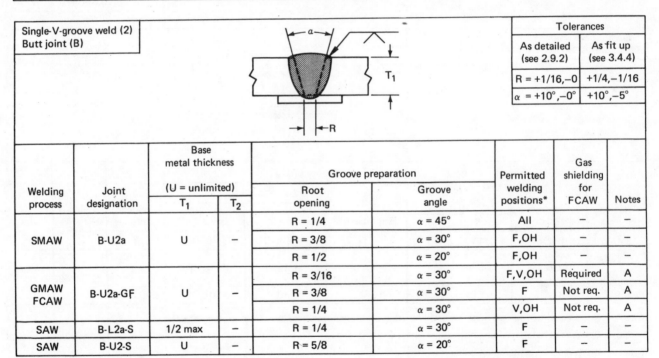

Square-groove weld (1)
T-joint (T)
Corner joint (C)

| Welding process | Joint designation | Base metal thickness (U = unlimited) | | Groove preparation | | | Permitted welding positions | Gas shielding for FCAW | Notes |
| | | T₁ | T₂ | Root opening | Tolerances | | | | |
					As detailed (see 2.9.2)	As fit up (see 3.3.4)			
SMAW	TC-L1b	1/4 max	U	$R = T_1/2$	+1/16,−0	+1/16,−1/8	All	−	C,J
GMAW FCAW	TC-L1-GF	3/8 max	U	R = 0 to 1/8	+1/16,−0	+1/16,−1/8	All	Not req.	A,C,J
SAW	TC-L1-S	3/8 max	U	R = 0	±0	+1/16,−0	Flat	−	J

Single-V-groove weld (2)
Butt joint (B)

| Tolerances | |
As detailed (see 2.9.2)	As fit up (see 3.4.4)
R = +1/16,−0	+1/4,−1/16
α = +10°,−0°	+10°,−5°

| Welding process | Joint designation | Base metal thickness (U = unlimited) | | Groove preparation | | Permitted welding positions* | Gas shielding for FCAW | Notes |
		T₁	T₂	Root opening	Groove angle			
SMAW	B-U2a	U	−	R = 1/4	α = 45°	All	−	−
				R = 3/8	α = 30°	F,OH	−	−
				R = 1/2	α = 20°	F,OH	−	−
GMAW FCAW	B-U2a-GF	U	−	R = 3/16	α = 30°	F,V,OH	Required	A
				R = 3/8	α = 30°	F	Not req.	A
				R = 1/4	α = 30°	V,OH	Not req.	A
SAW	B-L2a-S	1/2 max	−	R = 1/4	α = 30°	F	−	−
SAW	B-U2-S	U	−	R = 5/8	α = 20°	F	−	−

Note A: Not prequalified for gas metal arc welding using short circuiting transfer. Refer to Appendix D.

Note C: Gouge root before welding other side.

Note J: If fillet welds are used in buildings to reinforce groove welds in corner and T-joints, they shall be equal to 1/4 T_1 but need not exceed 3/8 in. Groove welds in corner and T-joints of bridges shall be reinforced with fillet welds equal to 1/4 T_1 but not more than 3/8 in.

* F = Flat, OH = Overhead.

Fig. 2.9.1 (continued)—Prequalified complete joint penetration groove welded joints

Single-V-groove weld (2)
Corner joint (B)

Tolerances		
	As detailed (see 2.9.2)	As fit up (see 3.3.4)
R =	+1/16,−0	+1/4,−1/16
α =	+10°,−0°	+10°,−5°

Welding process	Joint designation	Base metal thickness U = unlimited		Groove preparation		Permitted welding positions	Gas shielding for FCAW	Notes
		T_1	T_2	Root opening	Groove angle			
SMAW	C-U2a	U	U	R = 1/4	α = 45°	All	−	−
				R = 3/8	α = 30°	F,OH	−	−
GMAW FCAW	C-U2a-GF	U	U	R = 1/2	α = 20°	F,OH	−	−
				R = 3/16	α = 30°	F,V,OH	Required	A
				R = 3/8	α = 30°	F	Not req.	A
				R = 1/4	α = 30°	V,OH	Not req.	A
SAW	C-L2a-S	1/2 max	U	R = 1/4	α = 30°	F	−	−
SAW	C-U2-S	U	U	R = 5/8	α = 20°	F	−	−

Single-V-groove weld (2)
Butt joint (B)

Welding process	Joint designation	Base metal thickness (U = unlimited)		Groove preparation			Permitted welding positions	Gas shielding for FCAW	Notes
		T_1	T_2	Root opening Root face Groove angle	Tolerances				
					As detailed (see 2.9.2)	As fit up (see 3.3.4)			
SMAW	B-U2	U	−	R = 0 to 1/8 f = 0 to 1/8 α = 60°	+1/16,−0 +1/16,−0 +10°,−0°	+1/16,−1/8 Not limited +10°,−5°	All	−	C
GMAW FCAW	B-U2-GF	U	−	R = 0 to 1/8 f = 0 to 1/8 α = 60°	+1/16,−0 +1/16,−0 +10°,−0°	+1/16,−1/8 Not limited +10°,−5°	All	Not required	A,C
SAW	B-L2b-S	Over 1/2 to 1 inclusive	−	R = 0 f = 1/4 max α = 60°	±0 +0,−1/4 +10°,−0°	+1/16,−0 ±1/16 +10°,−5°	Flat	−	K
SAW	B-L2c-S	Over 1/2 to 1	−	R = 0, α = 60° f = 1/4 max	R = ±0 f = +0,−f α = +10°,−0°	+1/16,−0 ±1/16 +10°,−5°	Flat	−	C
		Over 1 to 1-1/2	−	R = 0, α = 60° f = 1/2 max					
		Over 1-1/2 to 2	−	R = 0, α = 60°. f = 5/8 max					

Note A: Not prequalified for gas metal arc welding using short circuiting transfer. Refer to Appendix D.

Note C: Gouge root before welding other side.

Note K: Weld root after welding at least one pass on arrow side.

Fig. 2.9.1 (continued)—Prequalified complete joint penetration groove welded joints

Single-V-groove weld (2)
Corner joint (C)

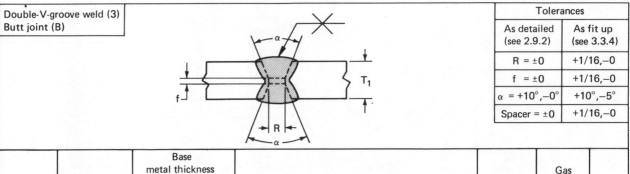

Welding process	Joint designation	Base metal thickness (U = unlimited)		Groove preparation			Permitted welding positions	Gas shielding for FCAW	Notes
		T_1	T_2	Root opening Root face Groove angle	Tolerances As detailed (see 2.9.2)	As fit up (see 3.3.4)			
SMAW	C-U2	U	U	R = 0 to 1/8 f = 0 to 1/8 α = 60°	+1/16,−0 +1/16,−0 +10°,−0°	+1/16,−1/8 Not limited +10°,−5°	All	−	C,J
GMAW FCAW	C-U2-GF	U	U	R = 0 to 1/8 f = 0 to 1/8 α = 60°	+1/16,−0 +1/16,−0 +10°,−0°	+1/16,−1/8 Not limited +10°,−5°	All	Not required	A,C,J
SAW	C-L2b-S	1 max	U	R = 0 f = 1/4 max α = 60°	±0 +0,−1/4 +10°,−0°	+1/16,−0 ±1/16 +10°,−5°	Flat	−	J,K

Double-V-groove weld (3)
Butt joint (B)

	Tolerances	
	As detailed (see 2.9.2)	As fit up (see 3.3.4)
R =	±0	+1/16,−0
f =	±0	+1/16,−0
α =	+10°,−0°	+10°,−5°
Spacer =	±0	+1/16,−0

Welding process	Joint designation	Base metal thickness (U = unlimited)		Groove preparation			Permitted welding positions*	Gas shielding for FCAW	Notes
		T_1	T_2	Root opening	Root face	Groove angle			
SMAW	B-U3a	U, preferably 5/8 or thicker Spacer = 1/8 x R	−	R = 1/4	f = 0 to 1/8	α = 45°	All	−	C,M
				R = 3/8	f = 0 to 1/8	α = 30°	F,OH	−	
				R = 1/2	f = 0 to 1/8	α = 20°	F,OH	−	
SAW	B-U3a-S	U Spacer = 1/4 x R	−	R = 5/8	f = 0 to 1/4	α = 20°	F	−	M

Note A: Not prequalified for gas metal arc welding using short circuiting transfer. Refer to Appendix D.

Note C: Gouge root before welding other side.

Note J: If fillet welds are used in buildings to reinforce groove welds in corner and T-joints, they shall be equal to 1/4 T_1 but need not exceed 3/8 in. Groove welds in corner and T-joints of bridges shall be reinforced with fillet welds equal to 1/4 T_1 but not more than 3/8 in.

Note K: Weld root after welding at least one pass on arrow side.

Note M: Double-groove welds may have grooves of unequal depth, but the depth of the shallower groove shall be no less than one-fourth of the thickness of the thinner part joined.

    * F = Flat, OH = Overhead.

Fig. 2.9.1 (continued)—Prequalified complete joint penetration groove welded joints

Double-V-groove weld (3) Butt joint (B)					

For B-U3c-S only

	T_1	S_1
Over	to	
2	2-1/2	1-3/8
2-1/2	3	1-3/4
3	3-5/8	2-1/8
3-5/8	4	2-3/8
4	4-3/4	2-3/4
4-3/4	5-1/2	3-1/4
5-1/2	6-1/4	3-3/4

For $T_1 > 6$-1/4, or $T_1 \leqslant 2$
$S_1 = 2/3 (T_1 - 1/4)$

Welding process	Joint designation	Base metal thickness (U = unlimited) T_1	T_2	Groove preparation Root opening Root face Groove angle	Tolerances As detailed (see 2.9.2)	Tolerances As fit up (see 3.3.4)	Permitted welding positions	Gas shielding for FCAW	Notes
SMAW	B-U3b	U, preferably 5/8 or thicker	–	R = 0 to 1/8 f = 0 to 1/8 $\alpha = \beta = 60°$	+1/16,–0 +1/16,–0 +10°,–0	+1/16,–1/8 Not limited +10°,–5°	All	–	C,M
GMAW FCAW	B-U3-GF						All	Not required	A,C,M
SAW	B-U3b-S	1-1/8 min	–	R = 1/8 f = 0 $\alpha = \beta = 60°$ $S_1 = 2/3\ T_1, S_2 = 3/8$ min	+1/16,–0 +1/16,–0 +10°,–0°	+1/16,–1/8 Not limited +10°,–5°	Flat	–	P
SAW	B-L3-S	1-1/2 max	–	R = 0 f = 1/4 max $\alpha = 60°; \beta = 80°$ $S_1 = 2/3 (T_1 - 1/4), S_2 = 1/3 (T_1 - 1/4)$	±0 +0,–1/4 +10°,–0°	+1/16,–0 Not limited +10°,–5°	Flat	–	K
SAW	B-U3c-S	U	–	R = 0 f = 1/4 max $\alpha = \beta = 60°$ To find S_1 see table above; $S_2 = (T_1 - S_1 + 1/4)$	±0 +0,–1/4 +10°,–0°	+1/16,–0 Not limited +10°,–5°	Flat	–	C

Note A: Not prequalified for gas metal arc welding using short circuiting transfer. Refer to Appendix D.

Note C: Gouge root before welding other side.

Note K: Weld root after welding at least one pass on arrow side.

Note M: Double-groove welds may have grooves of unequal depth, but the depth of the shallower groove shall be no less than one-fourth of the thickness of the thinner part joined.

Note P: Weld S_2 first with shielded metal arc low-hydrogen electrodes. Root of this weld shall be back gouged. Weld S_1 with single- or multiple-pass submerged arc weld in flat position after manual arc welding is completed on other side.

Fig. 2.9.1 (continued)—Prequalified complete joint penetration groove welded joints

Single-level-groove weld (4)
Butt joint (B)

	Tolerances	
	As detailed (see 2.9.2)	As fit up (see 3.3.4)
R =	+1/16,−0	+1/4,−1/16
α =	+10°,−0°	+10°,−5°

Welding process	Joint designation	Base metal thickness (U = unlimited)		Groove preparation		Permitted welding positions*	Gas shielding for FCAW	Notes
		T₁	T₂	Root opening	Groove angle			
SMAW	B-U4a	U	−	R = 1/4	α = 45°	All	---	Br
				R = 3/8	α = 30°	F,OH	−	Br
GMAW FCAW	B-U4a-GF	U	−	R = 3/16	α = 30°	All	Required	A,Br
				R = 1/4	α = 45°	All	Required	A,Br
				R = 3/8	α = 30°	Flat	Not req.	A,Br
				R = 1/4	α = 45°	All	Not req.	A,Br

Single-bevel-groove weld (4)
T-joint (T)
Corner joint (C)

	Tolerances	
	As detailed (see 2.9.2)	As fit up (see 3.3.4)
R =	+1/16,−0	+1/4,−1/16
α =	+10°,−0°	+10°,−5°

Welding process	Joint designation	Base metal thickness (U = unlimited)		Groove preparation		Permitted welding positions*	Gas shielding for FCAW	Notes
		T₁	T₂	Root opening	Groove angle			
SMAW	TC-U4c	U	U	R = 1/4	α = 45°	All	−	J,V
				R = 3/8	α = 30°	F, OH	−	J,V
GMAW FCAW	TC-U4c-GF	U	U	R = 3/16	α = 30°	All	Required	A, J, V
				R = 1/4	α = 45°	All	Required	A, J, V
				R = 3/8	α = 30°	Flat	Not req.	A, J, V
				R = 1/4	α = 45°	All	Not req.	A, J, V
SAW	TC-U4a-S	U	U	R = 3/8	α = 30°	Flat	−	J,V
				R = 1/4	α = 45°			

Note A: Not prequalified for gas metal arc welding using short circuiting transfer. Refer to Appendix D.

Note Br: Bridge application limits the use of these joints to the horizontal position (see 9.12.1.5).

Note J: If fillet welds are used in buildings to reinforce groove welds in corner and T-joints, they shall be equal to 1/4 T₁ but need not exceed 3/8 in. Groove welds in corner and T-joints of bridges shall be reinforced with fillet welds equal to 1/4 T₁ but not more than 3/8 in.

Note V: For corner joints, the outside groove preparation may be in either or both members, provided the basic groove configuration is not changed and adequate edge distance is maintained to support the welding operations without excessive edge melting.

*F = Flat, OH = Overhead.

Fig. 2.9.1 (continued)—Prequalified complete joint penetration groove welded joints

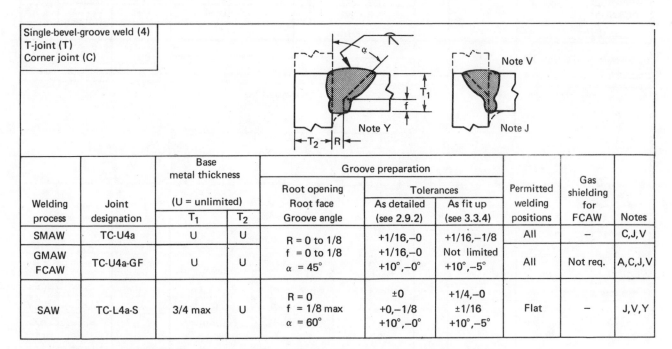

Welding process	Joint designation	Base metal thickness (U = unlimited) T_1	T_2	Groove preparation Root opening Root face Groove angle	Tolerances As detailed (see 2.9.2)	As fit up (see 3.3.4)	Permitted welding positions	Gas shielding for FCAW	Notes
SMAW	B-U4	U	–	R = 0 to 1/8	+1/16,–0	+1/16,–1/8	All	–	C
GMAW FCAW	B-U4-GF	U	–	f = 0 to 1/8 α = 45°	+1/16,–0 +10°,–0°	not limited +10°,–5°	All	Required	A,C

Welding process	Joint designation	Base metal thickness (U = unlimited) T_1	T_2	Groove preparation Root opening Root face Groove angle	Tolerances As detailed (see 2.9.2)	As fit up (see 3.3.4)	Permitted welding positions	Gas shielding for FCAW	Notes
SMAW	TC-U4a	U	U	R = 0 to 1/8	+1/16,–0	+1/16,–1/8	All	–	C,J,V
GMAW FCAW	TC-U4a-GF	U	U	f = 0 to 1/8 α = 45°	+1/16,–0 +10°,–0°	Not limited +10°,–5°	All	Not req.	A,C,J,V
SAW	TC-L4a-S	3/4 max	U	R = 0 f = 1/8 max α = 60°	±0 +0,–1/8 +10°,–0°	+1/4,–0 ±1/16 +10°,–5°	Flat	–	J,V,Y

Note A: Not prequalified for gas metal arc welding using short circuiting transfer. Refer to Appendix D.

Note C: Gouge root of joint before welding the other side.

Note J: If fillet welds are used in buildings to reinforce groove welds in corner and T-joints, they shall be equal to 1/4 T_1 but need not exceed 3/8 in. Groove welds in corner and T-joints of bridges shall be reinforced with fillet welds equal to 1/4 T_1 but not more than 3/8 in.

Note V: For corner joints, the outside groove preparation may be in either or both members, provided the basic groove configuration is not changed and adequate edge distance is maintained to support the welding operations without excessive edge melting.

Note Y: Shielded metal arc or submerged arc backing fillet weld required.

Fig. 2.9.1 (continued)—Prequalified complete joint penetration groove welded joints

Single-bevel-groove weld (4)
T-joint (T)
Corner joint (C)

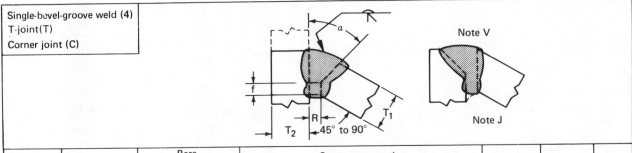

		Tolerances	
		As detailed (see 2.9.2)	As fit up (see 3.3.4)
		R = +1/16,−0	+1/4,−1/16
		α = +10°,−0°	+10°,−5°

Welding process	Joint designation	Base metal thickness (U = unlimited) T₁	T₂	Groove preparation Root opening	Groove angle	Permitted welding positions*	Gas shielding for FCAW	Notes
SMAW	TC-U4d	U	U	R = 1/4	$\alpha = 45°$	All	—	J,V
				R = 3/8	$\alpha = 30°$	F,OH	—	
GMAW FCAW	TC-U4d-GF	U	U	R = 3/16	$\alpha = 30°$	All	Required	A,J,V
				R = 1/4	$\alpha = 45°$	All		
				R = 3/8	$\alpha = 30°$	Flat	Not req.	
				R = 1/4	$\alpha = 45°$	All		
SAW	TC-U4b-S	U	U	R = 3/8	$\alpha = 30°$	Flat	—	J,V
				R = 1/4	$\alpha = 45°$			

Single-bevel-groove weld (4)
T-joint (T)
Corner joint (C)

Welding process	Joint designation	Base metal thickness (U = unlimited) T₁	T₂	Groove preparation Root opening / Root face / Groove angle	Tolerances As detailed (see 2.9.2)	As fit up (see 3.3.4)	Permitted welding positions	Gas shielding for FCAW	Notes
SMAW	TC-U4b	U	U	R = 0 to 1/8	+1/16,−0	+1/16,−1/8	All	—	C,J,V
GMAW FCAW	TC-U4b-GF	U	U	f = 0 to 1/8 α = 45°	+1/16,−0 +10°,−0°	Not limited +10°,−5°	All	Not required	A,C J,V
SAW	TC-L4b-S	3/4 max	U	R = 0 f = 1/8 max α = 60°	±0 +0,−1/8 +10°,−0°	+1/4,−0 ±1/16 +10°,−5°	Flat	—	J, V, Y

Note A: Not prequalified for gas metal arc welding using short circuiting transfer. Refer to Appendix D.

Note C: Gouge root of joint before welding the other side.

Note J: If fillet welds are used in buildings to reinforce groove welds in corner and T-joints, they shall be equal to 1/4 T₁ but need not exceed 3/8 in. Groove welds in corner and T-joints of bridges shall be reinforced with fillet welds equal to 1/4 T₁ but not more than 3/8 in.

Note V: For corner joints, the outside groove preparation may be in either or both members, provided the basic groove configuration is not changed and adequate edge distance is maintained to support the welding operations without excessive edge melting.

Note Y: Shielded metal arc or submerged arc backing weld required.

* F = Flat, OH = Overhead.

Fig. 2.9.1 (continued)—Prequalified complete joint penetration groove welded joints

Double-bevel-groove weld (5)									
Butt joint (B)									
T-joint (T)									
Corner joint (C)									

Tolerances	
As detailed (see 2.9.2)	As fit up (see 3.3.4)
R = ±0	+1/16,−0
f = +1/16,−0	±1/16
α = +10°,−0°	+10°,−5°
Spacer = ±0	+1/16,−0

Welding process	Joint designation	Base metal thickness (U = unlimited) T₁	T₂	Groove preparation Root opening	Root face	Groove angle	Permitted welding positions*	Gas shielding for FCAW	Notes
SMAW	B-U5b	U, preferably 5/8 or thicker Spacer = 1/8 X R	–	R = 1/4	f = 0 to 1/8	α = 45°	All	–	Br, C, M
	TC-U5a	U, preferably 5/8 or thicker Spacer = 1/8 X R	U	R = 1/4	f = 0 to 1/8	α = 45°	All	–	C, J, M, V
				R = 3/8	f = 0 to 1/8	α = 30°	F, OH	–	C, J, M, V

Note Br: Bridge application limits the use of these joints to the horizontal position (see 9.12.1.5).

Note C: Gouge root of joint before welding the other side.

Note J: If fillet welds are used in buildings to reinforce groove welds in corner and T-joints, they shall be equal to 1/4 T₁ but need not exceed 3/8 in. Groove welds in corner and T-joints of bridges shall be reinforced with fillet welds equal to 1/4 T₁ but not more than 3/8 in.

Note M: Double-groove welds may have grooves of unequal depth, but the depth of the shallower groove shall be no less than one-fourth of the thickness of the thinner part joined.

Note V: For corner joints, the outside groove preparation may be in either or both members, provided the basic groove configuration is not changed and adequate edge distance is maintained to support the welding operations without excessive edge melting.

*F = Flat, OH = Overhead.

Fig. 2.9.1 (continued)—Prequalified complete joint penetration groove welded joints

Double-bevel-groove weld (5)
Butt joint (B)

Limitations
Bridge applications limited to horizontal position (see 9.12.1.5).

Welding process	Joint designation	Base metal thickness (U = unlimited)		Groove preparation			Permitted welding positions	Gas shielding for FCAW	Notes
		T_1	T_2	Root opening Root face Groove angles	Tolerances As detailed (see 2.9.2)	As fit up (see 3.3.4)			
SMAW	B-U5a	U, preferably 5/8 or thicker	–	R = 0 to 1/8 f = 0 to 1/8 α = 45° β = 0° to 15°	+1/16,–0 +1/16,–0 $\alpha + \beta$, $^{+10}_{-0}$	+1/16,–1/8 Not limited $\alpha + \beta$, $^{+10}_{-5}$	All	–	C, M, Z
GMAW FCAW	B-U5-GF	U, preferably 5/8 or thicker	–	R = 0 to 1/8 f = 0 to 1/8 α = 45° β = 0°	+1/16,–0 +1/16,–0 +10°,–0° ±0°	+1/16,–1/8 Not limited +10°,–5° –	All	Not req.	A, C, M

Double-bevel-groove weld (5)
T-joint (T)
Corner joint (C)

Note V

Note J

Welding process	Joint designation	Base metal thickness (U = unlimited)		Groove preparation			Permitted welding positions	Gas shielding for FCAW	Notes
		T_1	T_2	Root opening Root face Groove angle	Tolerances As detailed (see 2.9.2)	As fit up (see 3.3.4)			
SMAW	TC-U5b	U, preferably 5/8 or thicker	U	R = 0 to 1/8 f = 0 to 1/8 α = 45°	+1/16,–0 +1/16,–0 +10°,–0°	+1/16,–1/8 Not limited +10°,–5°	All	–	C, J, M, V
GMAW FCAW	TC-U5-GF	U, preferably 5/8 or thicker	U				All	Not required	A, C, J, M, V
SAW	TC-U5-S	U	U	R = 0 f = 3/16 max α = 60°	±0 +0,–3/16 +10°,–0°	+1/16,–0 ±1/16 +10°,–5°	Flat	–	J, M, V

Note A: Not prequalified for gas metal arc welding using short circuiting transfer. Refer to Appendix D.

Note C: Gouge root of joint before welding the other side.

Note J: If fillet welds are used in buildings to reinforce groove welds in corner and T-joints, they shall be equal to 1/4 T_1 but need not exceed 3/8 in. Groove welds in corner and T-joints of bridges shall be reinforced with fillet welds equal to 1/4 T_1 but not more than 3/8 in.

Note M: Double-groove welds may have grooves of unequal depth, but the depth of the shallower groove shall be no less than one-fourth of the thickness of the thinner part joined.

Note V: For corner joints, the outside groove preparation may be in either or both members, provided the basic groove configuration is not changed and adequate edge distance is maintained to support the welding operations without excessive edge melting.

Note Z: When lower plate is beveled, make the first root pass on this side.

Fig. 2.9.1 (continued)—Prequalified complete joint penetration groove welded joints

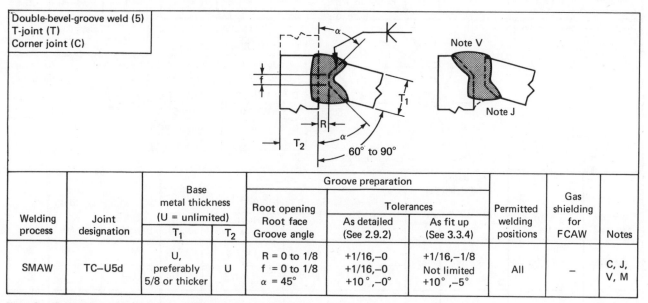

Double-bevel-groove weld (5)
T-joint (T)
Corner joint (C)

	Tolerances	
	As detailed (see 2.9.2)	As fit up (see 3.3.4)
R = ±0	+1/16,−0	
f = +1/16,−0	Not limited	
α = +10°,−0°	+10°,−5°	
Spacer = ±0	+1/16,−0	

Welding process	Joint designation	Base metal thickness (U = unlimited)		Groove preparation			Permitted welding positions*	Gas shielding for (FCAW)	Notes
		T_1	T_2	Root opening	Root face	Groove angle			
SMAW	TC-U5c	U, preferably 5/8 or thicker Spacer = 1/8 X R	U	R = 1/4	f = 0 to 1/8	α = 45°	All	–	C, J, V, M
				R = 3/8	f = 0 to 1/8	α = 30°	F, OH	–	C, J, V, M

Double-bevel-groove weld (5)
T-joint (T)
Corner joint (C)

Welding process	Joint designation	Base metal thickness (U = unlimited)		Groove preparation			Permitted welding positions	Gas shielding for FCAW	Notes
		T_1	T_2	Root opening Root face Groove angle	Tolerances As detailed (See 2.9.2)	As fit up (See 3.3.4)			
SMAW	TC–U5d	U, preferably 5/8 or thicker	U	R = 0 to 1/8 f = 0 to 1/8 α = 45°	+1/16,−0 +1/16,−0 +10°,−0°	+1/16,−1/8 Not limited +10°,−5°	All	–	C, J, V, M

Note C: Gouge root of joint before welding the other side.

Note J: If fillet welds are used in buildings to reinforce groove welds in corner and T-joints, they shall be equal to 1/4 T_1 but need not exceed 3/8 in. Groove welds in corner and T-joints of bridges shall be reinforced with fillet welds equal to 1/4 T_1 but not more than 3/8 in.

Note M: Double-groove welds may have grooves of unequal depth, but the depth of the shallower groove shall be no less than one-fourth of the thickness of the thinner part joined.

Note V: For corner joints, the outside groove preparation may be in either or both members, provided the basic groove configuration is not changed and adequate edge distance is maintained to support the welding operations without excessive edge melting.

*F = Flat, OH = Overhead.

Fig. 2.9.1 (continued)—Prequalified complete joint penetration groove welded joints

Single-U-groove weld (6)
Butt joint (B)
Corner joint (C)

Tolerances		
	As detailed (see 2.9.2)	As fit up (see 3.3.4)
R =	+1/16,−0	+1/4,−1/16
α =	+10°,−0°	+10°,−5°
f =	±1/16	Not limited
r =	+1/4,−0	±1/16

Welding process	Joint designation	Base metal thickness (U = unlimited)		Groove preparation				Permitted welding positions*	Gas shielding for FCAW	Notes
		T₁	T₂	Root opening	Groove angle	Root face	Groove radius			
SMAW	B-U6	U	U	R = 0° to 1/8	α = 45°	f = 1/8	r = 1/4	All	—	C
				R = 0° to 1/8	α = 20°	f = 1/8	r = 1/4	F,OH	—	C
	C-U6	U	U	R = 0° to 1/8	α = 45°	f = 1/8	r = 1/4	All	—	C,J
				R = 0° to 1/8	α = 20°	f = 1/8	r = 1/4	F,OH	—	C,J
GMAW FCAW	B-U6-GF	U	U	R = 0° to 1/8	α = 20°	f = 1/8	r = 1/4	All	Not req.	A,C
	C-U6-GF	U	U	R = 0° to 1/8	α = 20°	f = 1/8	r = 1/4	All	Not req.	A,C,J

Double-V-groove weld (7)
Butt joint (B)

Tolerances					
For B-U7 and B-U7-GF			For B-U7-S		
	As detailed (see 2.9.2)	As fit up (see 3.3.4)		As detailed (see 2.9.2)	As fit up (see 3.3.4)
R =	+1/16,−0	+1/16,−1/8	R =	±0	+1/16,−0
α =	+10°,−0	+10°,−5°	f =	+0,−1/4	±1/16
f =	+1/16,−0	Not limited			
r =	+1/4,−0	±1/16			

Welding process	Joint designation	Base metal thickness (U = unlimited)		Groove preparation				Permitted welding positions*	Gas shielding for FCAW	Notes
		T₁	T₂	Root opening	Groove angle	Root face	Groove radius			
SMAW	B-U7	U, preferably 5/8 or thicker	—	R = 0 to 1/8	α = 45°	f = 1/8	r = 1/4	All	—	C
				R = 0 to 1/8	α = 20°	f = 1/8	r = 1/4	F,OH	—	C
GMAW FCAW	B-U7-GF	U, preferably 5/8 or thicker	—	R = 0 to 1/8	α = 45°	f = 1/8	r = 1/4	All	Not required	A,C
SAW	B-U7-S	U	—	R = 0	α = 20°	f = 1/4 max	r = 1/4	F	—	—

Note A: Not prequalified for gas metal arc welding using short circuiting transfer. Refer to Appendix D.

Note C: Gouge root of joint before welding the other side.

Note J: If fillet welds are used in buildings to reinforce groove welds in corner and T-joints, they shall be equal to 1/4 T₁ but need not exceed 3/8 in. Groove welds in corner and T-joints of bridges shall be reinforced with fillet welds equal to 1/4 T₁ but not more than 3/8 in.

Note M: Double-groove welds may have grooves of unequal depth, but the depth of the shallower groove shall be no less than one-fourth of the thickness of the thinner part joined.

* F = Flat, OH = Overhead.

Fig. 2.9.1 (continued)—Prequalified complete joint penetration groove welded joints

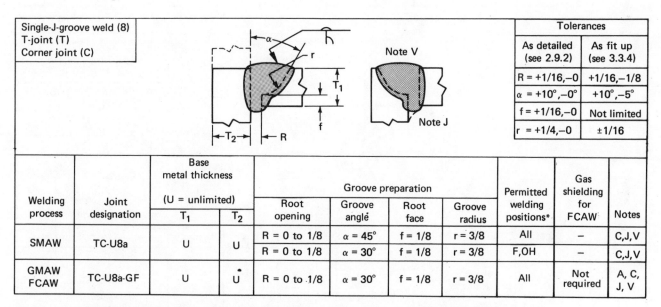

Welding process	Joint designation	Base metal thickness (U = unlimited)		Groove preparation				Permitted welding positions	Gas shielding for FCAW	Notes
		T_1	T_2	Root opening	Groove angle	Root face	Groove radius			
SMAW	B-U8	U	–	R = 0 to 1/8	α = 45°	f = 1/8	r = 3/8	All	–	Br, C
GMAW FCAW	B-U8-GF	U	–	R = 0 to 1/8	α = 30°	f = 1/8	r = 3/8	All	Not required	A,Br,C

Welding process	Joint designation	Base metal thickness (U = unlimited)		Groove preparation				Permitted welding positions*	Gas shielding for FCAW	Notes
		T_1	T_2	Root opening	Groove angle	Root face	Groove radius			
SMAW	TC-U8a	U	U	R = 0 to 1/8	α = 45°	f = 1/8	r = 3/8	All	–	C,J,V
				R = 0 to 1/8	α = 30°	f = 1/8	r = 3/8	F,OH	–	C,J,V
GMAW FCAW	TC-U8a-GF	U	U	R = 0 to 1/8	α = 30°	f = 1/8	r = 3/8	All	Not required	A, C, J, V

Note A: Not prequalified for gas metal arc welding using short circuiting transfer. Refer to Appendix D.

Note Br: Bridge application limits the use of these joints to the horizontal position (see 9.12.1.5).

Note C: Gouge root before welding other side.

Note J: If fillet welds are used in buildings to reinforce groove welds in corner and T-joints, they shall be equal to 1/4 T_1 but need not exceed 3/8 in. Groove welds in corner and T-joints of bridges shall be reinforced with fillet welds equal to 1/4 T_1 but not more than 3/8 in.

Note V: For corner joints, the outside groove preparation may be in either or both members, provided the basic groove configuration is not changed and adequate edge distance is maintained to support the welding operations without excessive edge melting.

* F = Flat, OH = Overhead.

Fig. 2.9.1 (continued)—Prequalified complete joint penetration groove welded joints

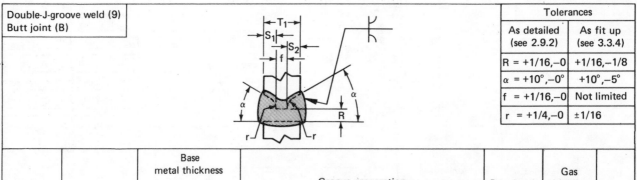

Single-J-groove weld (8)
T-joint (T)
Corner joint (C)

Note V

Note J

45° to 90°

	Tolerances	
	As detailed (see 2.9.2)	As fit up (see 3.3.4)
R =	+1/16,−0	+1/16,−1/8
α =	+10°,−0°	+10°,−5°
f =	+1/16,−0	Not limited
r =	+1/4,−0	±1/16

Welding process	Joint designation	Base metal thickness (U = unlimited)		Groove preparation				Permitted welding positions*	Gas shielding for FCAW	Notes
		T₁	T₂	Root opening	Groove angle	Root face	Groove radius			
SMAW	TC-U8b	U	U	R = 0 to 1/8	α = 45°	f = 1/8	r = 3/8	All	−	C,J,V
				R = 0 to 1/8	α = 30°	f = 1/8	r = 3/8	F,OH	−	C,J,V
GMAW FCAW	TC-U8b-GF	U	U	R = 0 to 1/8	α = 30°	f = 1/8	r = 3/8	All	Not required	A,C, J,V

Double-J-groove weld (9)
Butt joint (B)

	Tolerances	
	As detailed (see 2.9.2)	As fit up (see 3.3.4)
R =	+1/16,−0	+1/16,−1/8
α =	+10°,−0°	+10°,−5°
f =	+1/16,−0	Not limited
r =	+1/4,−0	±1/16

Welding process	Joint designation	Base metal thickness (U = unlimited)		Groove preparation				Permitted welding positions	Gas shielding for FCAW	Notes
		T₁	T₂	Root opening	Groove angle	Root face	Groove radius			
SMAW	B-U9	U, preferably 5/8 or thicker	−	R = 0 to 1/8	α = 45°	f = 1/8	r = 3/8	All	−	Br, C, M
GMAW FCAW	B-U-GF	U, preferably 5/8 or thicker	−	R = 0 to 1/8	α = 30°	f = 1/8	r = 3/8	All	Not required	A, Br, C, M

Note A: Not prequalified for gas metal arc welding using short circuiting transfer. Refer to Appendix D.

Note Br: Bridge application limits the use of these joints to the horizontal position (see 9.12.1.5).

Note C: Gouge root before welding other side.

Note J: If fillet welds are used in buildings to reinforce groove welds in corner and T-joints, they shall be equal to 1/4 T₁ but need not exceed 3/8 in. Groove welds in corner and T-joints of bridges shall be reinforced with fillet welds equal to 1/4 T₁ but not more than 3/8 in.

Note M: Double-groove welds may have grooves of unequal depth, but the depth of the shallower groove shall be no less than one-fourth of the thickness of the thinner part joined.

Note V: For corner joints, the outside groove preparation may be in either or both members, provided the basic groove configuration is not changed and adequate edge distance is maintained to support the welding operations without excessive edge melting.

* F = Flat, OH = Overhead.

Fig. 2.9.1 (continued)—Prequalified complete joint penetration groove welded joints

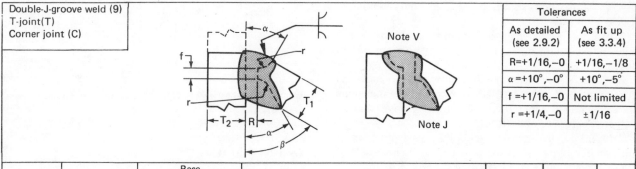

Double-J-groove weld (9)
T-joint (T)
Corner joint (C)

Tolerances	
As detailed (see 2.9.2)	As fit up (see 3.3.4)
R = +1/16,−0	+1/16,−1/8
α = +10°,−0°	+10°,−5°
f = +1/16,−0	Not limited
r = +1/4,−0	±1/16

Welding process	Joint designation	Base metal thickness (U = unlimited)		Groove preparation				Permitted welding positions	Gas shielding for FCAW	Notes
		T₁	T₂	Root opening	Groove angle	Root face	Groove radius			
SMAW	TC-U9a	U, preferably 5/8 or thicker	U	R = 0 to 1/8	α = 45°	f = 1/8	r = 3/8	All	−	C, J, M, V
				R = 0 to 1/8	α = 30°	f = 1/8	r = 3/8	F,OH	−	C, J, M, V
GMAW FCAW	TC-U9a-GF	U, preferably 5/8 or thicker	U	R = 0 to 1/8	α = 30°	f = 1/8	r = 3/8	All	Not required	A, C, J, M, V

Double-J-groove weld (9)
T-joint(T)
Corner joint (C)

Tolerances	
As detailed (see 2.9.2)	As fit up (see 3.3.4)
R=+1/16,−0	+1/16,−1/8
α =+10°,−0°	+10°,−5°
f =+1/16,−0	Not limited
r =+1/4,−0	±1/16

Welding process	Joint designation	Base metal thickness (U = unlimited)		Groove preparation				Permitted welding positions*	Gas shielding for FCAW	Notes
		T₁	T₂	Root opening	Groove angle	Root face	Groove radius			
SMAW	TC-U9b	U, preferably 5/8 or thicker	U	R = 0 to 1/8	α = 45°	f = 1/8	r = 3/8	All	−	C, J, M, V
				R = 0 to 1/8	α = 30°	f = 1/8	r = 3/8	F,OH		
GMAW FCAW	TC-U9b-GF	U, preferably 5/8 or thicker	U	R = 0 to 1/8	α = 30°	f = 1/8	r = 3/8	All	Not required	A, C, J, M, V

Note A: Not prequalified for gas metal arc welding using short circuiting transfer. Refer to Appendix D.

Note C: Gouge root before welding other side.

Note J: If fillet welds are used in buildings to reinforce groove welds in corner and T-joints, they shall be equal to 1/4 T_1 but need not exceed 3/8 in. Groove welds in corner and T-joints of bridges shall be reinforced with fillet welds equal to 1/4 T_1 but not more than 3/8 in.

Note M: Double-groove welds may have grooves of unequal depth, but the depth of the shallower groove shall be no less than one-fourth of the thickness of the thinner part joined.

Note V: For corner joints, the outside groove preparation may be in either or both members, provided the basic groove configuration is not changed and adequate edge distance is maintained to support the welding operations without excessive edge melting.

    * F = Flat, OH = Overhead.

Fig. 2.9.1 (continued)—Prequalified complete joint penetration groove welded joints

2.10 Partial Joint Penetration Groove Welds

2.10.1 Partial joint penetration groove welds made by shielded metal arc welding, submerged arc welding, gas metal arc welding (except short circuiting transfer), or flux cored arc welding in butt, corner, and T-joints which may be used without performing the joint welding procedure qualification tests prescribed in 5.2 are detailed in Fig. 2.10.1 and are subject to the limitations specified in 2.10.2.

2.10.1.1 Definition. Except as provided in 10.13.1.1, groove welds without steel backing, welded from one side, and groove welds welded from both sides but without back gouging are considered partial joint penetration groove welds.

2.10.1.2 All partial joint penetration groove welds made by short circuiting transfer gas metal arc welding (see Appendix D) shall be qualified by the joint welding procedure qualification test prescribed in 5.2.

2.10.2 Dimensional Tolerances. Dimensions of groove welds specified on design or detailed drawings may vary from the dimensions shown in Fig. 2.10.1 only within the following limits.

2.10.2.1 The groove angle is minimum; it may be detailed to exceed the dimensions shown by no more than 10 degrees.

2.10.2.2 The radius of the J-grooves and U-grooves is minimum. It may be detailed to exceed the dimensions shown by no more than 1/8 in. (3 mm). U-grooves may be prepared before or after fit-up.

2.10.2.3 Double-groove welds may have grooves of unequal depth, providing the weld deposit on each side of the joint conforms to the limitations of Fig. 2.10.1.

2.10.2.4 The minimum root face of the joints shall be 1/8 in. (3 mm), except that the minimum root face for joints to be welded by submerged arc welding shall be 1/4 in. (6 mm).

2.10.3 The effective throat of partial joint penetration square-, single- or double-V-, bevel-, J-, and U-groove welds shall be as shown in Table 2.10.3.

2.10.3.1 Shop or working drawings shall specify the groove depths (S) applicable for the effective throat (E) required for the welding process and position of welding to be used.

2.10.4 Groove preparations detailed for prequalified shielded metal arc welded joints may be used for prequalified gas metal arc or flux cored arc welding.

2.10.5 Corner Joints. For corner joints the outside groove preparation may be in either or both members, provided the basic groove configuration is not changed and adequate edge distance is maintained to support the welding operations without excessive melting.

Table 2.10.3
Minimum effective throat for partial joint penetration groove welds

Base metal thickness of thicker part joined		Minimum effective throat	
in.	mm	in.	mm
To 1/4 (6.4) incl.		1/8*	3
Over 1/4 (6.4) to 1/2 (12.7) incl.		3/16	5
Over 1/2 (12.7) to 3/4 (19.0) incl.		1/4	6
Over 3/4 (19.0) to 1-1/2 (38.1) incl.		5/16	8
Over 1-1/2 (38.1) to 2-1/4 (57.1) incl.		3/8	10
Over 2-1/4 (57.1) to 6 (152) incl.		1/2	13
Over 6 (152)		5/8	16

*Minimum size for bridge applications is 3/16 in.

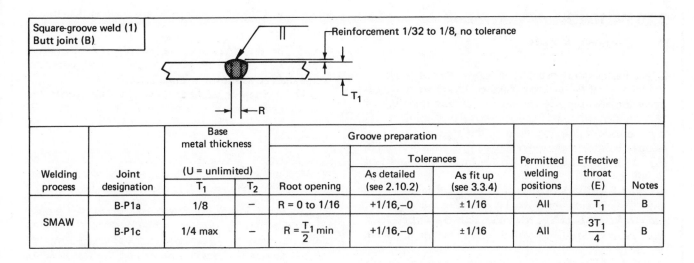

Welding process	Joint designation	Base metal thickness (U = unlimited)		Groove preparation			Permitted welding positions	Effective throat (E)	Notes
		T_1	T_2	Root opening	Tolerances				
					As detailed (see 2.10.2)	As fit up (see 3.3.4)			
SMAW	B-P1a	1/8	–	R = 0 to 1/16	+1/16,–0	±1/16	All	T_1	B
	B-P1c	1/4 max	–	R = $\frac{T}{2}$1 min	+1/16,–0	±1/16	All	$\frac{3T_1}{4}$	B

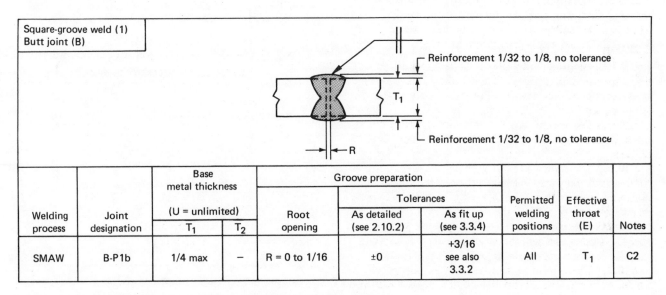

Welding process	Joint designation	Base metal thickness (U = unlimited)		Groove preparation			Permitted welding positions	Effective throat (E)	Notes
		T_1	T_2	Root opening	Tolerances				
					As detailed (see 2.10.2)	As fit up (see 3.3.4)			
SMAW	B-P1b	1/4 max	–	R = 0 to 1/16	±0	+3/16 see also 3.3.2	All	T_1	C2

Note B: Joints welded from one side. These welds are not applicable to bridges.

Note C2: Root need not be gouged before welding second side. This weld is not applicable to bridges.

Fig. 2.10.1—Prequalified partial joint penetration groove welded joints

Welding process	Joint designation	Base metal thickness (U = unlimited)		Groove preparation			Permitted welding positions	Effective throat (E)	Notes
		T_1	T_2	Root opening Root face Groove angle	Tolerances				
					As detailed (see 2.10.2)	As fit up (see 3.3.4)			
SMAW	B-P2	1/2 max	–	R = 3/32 min f = 0 to 1/8 α = 60°	±1/16 ±1/16 +10°,−0°	±1/16 ±1/16 +10°,−5°	All	T_1	B, L
SMAW	BC-P2	1/4 min (for bridges 5/16 min)	U	R = 0 f = 1/8 min α = 60°	±0 ±1/16 +10°,−0°	+1/16,−0 ±1/16 +10°,−5°	All	S	E, L
GMAW FCAW	BC-P2-GF	1/4 min (for bridges 5/16 min)	U	R = 0 f = 1/8 min α = 60°	±0 ±1/16 +10°,−0°	+1/16,−0 ±1/16 +10°,−5°	All	S	A, E, L
SAW	BC-P2-S	3/8 min (for bridges 7/16 min)	U	R = 0 f = 1/4 min α = 60°	±0 ±1/16 +10°,−0°	+1/16,−0 ±1/16 +10°,−5°	Flat	S	E, L

Note A: Not prequalified for gas metal arc welding using short circuiting transfer. Refer to Appendix D.

Note B: Joint is welded from one side only.

Note E: Minimum effective throat (E) as shown in Table 2.10.3; S as specified on drawings.

Note L: Butt and T-joints are not prequalified for bridges.

Fig. 2.10.1 (continued)—Prequalified partial joint penetration groove welded joints

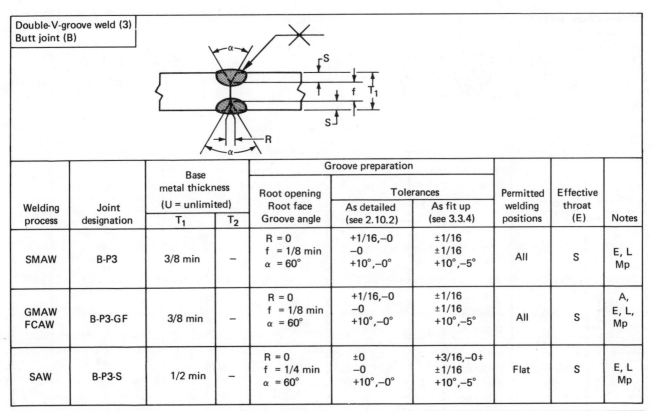

Double-V-groove weld (3)
Butt joint (B)

Welding process	Joint designation	Base metal thickness (U = unlimited)		Groove preparation			Permitted welding positions	Effective throat (E)	Notes
		T₁	T₂	Root opening Root face Groove angle	Tolerances				
					As detailed (see 2.10.2)	As fit up (see 3.3.4)			
SMAW	B-P3	3/8 min	–	R = 0 f = 1/8 min α = 60°	+1/16,–0 –0 +10°,–0°	±1/16 ±1/16 +10°,–5°	All	S	E, L Mp
GMAW FCAW	B-P3-GF	3/8 min	–	R = 0 f = 1/8 min α = 60°	+1/16,–0 –0 +10°,–0°	±1/16 ±1/16 +10°,–5°	All	S	A, E, L, Mp
SAW	B-P3-S	1/2 min	–	R = 0 f = 1/4 min α = 60°	±0 –0 +10°,–0°	+3/16,–0‡ ±1/16 +10°,–5°	Flat	S	E, L Mp

Single-level-groove weld (4)
Butt joint (B)

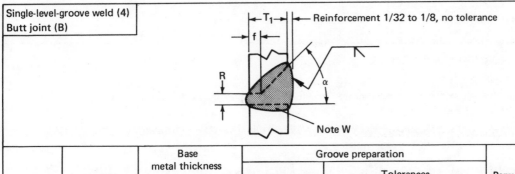

Reinforcement 1/32 to 1/8, no tolerance

Note W

Welding process	Joint designation	Base metal thickness (U = unlimited)		Groove preparation			Permitted welding positions	Effective throat (E)	Notes
		T₁	T₂	Root opening Root face Groove angle	Tolerances				
					As detailed (see 2.10.2)	As fit up (see 3.3.4)			
SMAW	B-P4	1/2 max	–	R = 3/32 min f = 0 to 1/8 α = 45°	R = min ±1/16 +10°,–0°	±1/16 ±1/16 +10°,–5°	All	T₁	B, L, W

Note A: Not prequalified for gas metal arc welding using short circuiting transfer. Refer to Appendix D.

Note B: Joint is welded from one side only.

Note E: Minimum effective throat (E) as shown in Table 2.10.3; S as specified on drawings.

Note L: Butt and T-joints are not prequalified for bridges.

Note Mp: Double-groove welds may have grooves of unequal depth, provided they conform to the limitations of Note E. Also, the effective throat (E), less any reduction, applies individually to each groove.

Note W: Unbeveled face is the lower edge for horizontal position.

‡ Fit-up tolerance, SAW: see 3.3.2; for rolled shapes R may be 5/16 inches in thick plates if backing is provided.

Fig. 2.10.1 (continued)—Prequalified partial joint penetration groove welded joints

| Single-bevel-groove (4) Butt joint (B) T-joint(T) Corner joint (C) | | | | | | | | | |

Welding process	Joint designation	Base metal thickness (U = unlimited)		Groove preparation			Permitted welding positions*	Effective throat (E)	Notes
		T_1	T_2	Root opening Root face Groove angle	Tolerances				
					As detailed (see 2.10.2)	As fit up (see 3.3.4)			
SMAW	BTC-P4	1/4 min (for bridges 5/16 min)	U	R = 0 f = 1/8 min α = 45°	+1/16,−0 −0 +10°,−0°	±1/16 ±1/16 +10°,−5°	All	S −1/8	E, L, V
GMAW FCAW	BTC-P4-GF	1/4 min (for bridges 5/16 min)	U	R = 0 f = 1/8 min α = 45°	+1/16,−0 −0 +10°,−0°	±1/16 ±1/16 +10°,−5°	F,H	S	A, E, L, V
							V,OH	S −1/8	
SAW	TC-P4-S	3/8 min (for bridges 7/16 min)	U	R = 0 f = 1/4 min α = 60°	±0 −0 +10°,−0°	+3/16,−0‡ ±1/16 +10°,−5°	Flat	S	E, L, V

Note A: Not prequalified for gas metal arc welding using short circuiting transfer. Refer to Appendix D.

Note E: Minimum effective throat (E) as shown in Table 2.10.3; S as specified on drawings.

Note L: Butt and T-joints are not prequalified for bridges.

Note V: For corner joints, the outside groove preparation may be in either or both members, provided the basic groove configuration is not changed and adequate edge distance is maintained to support the welding operations without excessive edge melting.

‡ Fit-up tolerance, SAW: see 3.3.2; for rolled shapes R may be 5/16 inches in thick plates if backing is provided.

* F = Flat, H = Horizontal, V = Vertical, OH = Overhead.

Fig. 2.10.1 (continued)—Prequalified partial joint penetration groove welded joints

| Double-bevel-groove weld (5) |
| Butt joint (B) |
| T-joint (T) |
| Corner joint (C) |

Welding process	Joint designation	Base metal thickness (U = unlimited)		Groove preparation				Permitted welding positions*	Effective throat (E)	Notes
		T₁	T₂	Root opening Root face Groove angle	Tolerances					
					As detailed (see 2.10.2)	As fit up (see 3.3.4)				
SMAW	BTC-P5	3/8 min (for bridges 1/2 min)	U	$R \simeq 0$ $f = 1/8$ min $\alpha = 45°$	+1/16,−0 −0 +10°,−0°	±1/16 ±1/16 +10°,−5°		All	S −1/8	E, L, Mp, V
GMAW FCAW	BTC-P5-GF	3/8 min (for bridges 1/2 min)	U	$R = 0$ $f = 1/8$ min $\alpha = 45°$	+1/16,−0 −0 +10°,−0°	±1/16 ±1/16 +10°,−5°		F,H	S	A, E, L, Mp, V
								V,OH	S −1/8	
SAW	TC-P5-S	1/2 min (for bridges 5/8 min)	U	$R = 0$ $f = 1/4$ min $\alpha = 60°$	±0 −0 +10°,−0°	+3/16,−0‡ ±1/16 +10°,−5°		Flat	S	E, L, Mp, V

Note A: Not prequalified for gas metal arc welding using short circuiting transfer. Refer to Appendix D.

Note E: Minimum effective throat (E) as shown in Table 2.10.3; S as specified on drawings.

Note L: Butt and T-joints are not prequalified for bridges.

Note Mp: Double-groove welds may have grooves of unequal depth, provided they conform to the limitations of Note E. Also, the effective throat (E), less any reduction, applies individually to each groove.

Note V: For corner joints, the outside groove preparation may be in either or both members, provided the basic groove configuration is not changed and adequate edge distance is maintained to support the welding operations without excessive edge melting.

‡ Fit-up tolerance, SAW: see 3.3.2; for rolled shapes R may be 5/16 inches in thick plates if backing is provided.

* F = Flat, H = Horizontal, V = Vertical, OH = Overhead.

Fig. 2.10.1 (continued)—Prequalified partial joint penetration groove welded joints

Welding process	Joint designation	Base metal thickness (U = unlimited)		Groove preparation			Permitted welding positions	Effective throat (E)	Notes
		T_1	T_2	Root opening Root face Groove radius Groove angle	Tolerances				
					As detailed (see 2.10.2)	As fit up (see 3.3.4)			
SMAW	BC-P6	1/4 min (for bridges 5/16 min)	U	R = 0 to 1/8 f = 1/8 min r = 1/4 α = 45°	+1/16,−0 −0 +1/4,−0 +10°,−0°	±1/16 ±1/16 ±1/16 +10°,−5°	All	S	E, L
GMAW FCAW	BC-P6-GF	1/4 min (for bridges 5/16 min)	U	R = 0 f = 1/8 min r = 1/4 α = 20°	+1/16,−0 −0 +1/4,−0 +10°,−0°	±1/16 ±1/16 ±1/16 +10°,−5°	All	S	A, E, L
SAW	BC-P6-S	3/8 min (for bridges 7/16 min)	U	R = 0 f = 1/4 min r = 1/4 α = 20°	±0 −0 +1/4,−0 +10°,−0°	+3/16,−0‡ ±1/16 ±1/16 +10°,−5°	Flat	S	E, L

Note A: Not prequalified for gas metal arc welding using short circuiting transfer. Refer to Appendix D.

Note E: Minimum effective throat (E) as shown in Table 2.10.3; S as specified on drawings.

Note L: Butt and T-joints are not prequalified for bridges.

‡ Fit-up tolerance, SAW: see 3.3.2; for rolled shapes R may be 5/16 inches in thick plates if backing is provided.

Fig. 2.10.1 (continued)—Prequalified partial joint penetration groove welded joints

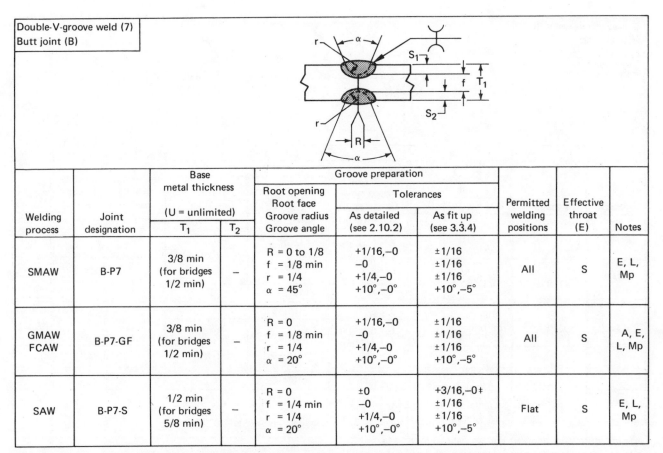

Welding process	Joint designation	Base metal thickness (U = unlimited)		Groove preparation			Permitted welding positions	Effective throat (E)	Notes
		T_1	T_2	Root opening Root face Groove radius Groove angle	Tolerances				
					As detailed (see 2.10.2)	As fit up (see 3.3.4)			
SMAW	B-P7	3/8 min (for bridges 1/2 min)	–	R = 0 to 1/8 f = 1/8 min r = 1/4 α = 45°	+1/16,–0 –0 +1/4,–0 +10°,–0°	±1/16 ±1/16 ±1/16 +10°,–5°	All	S	E, L, Mp
GMAW FCAW	B-P7-GF	3/8 min (for bridges 1/2 min)	–	R = 0 f = 1/8 min r = 1/4 α = 20°	+1/16,–0 –0 +1/4,–0 +10°,–0°	±1/16 ±1/16 ±1/16 +10°,–5°	All	S	A, E, L, Mp
SAW	B-P7-S	1/2 min (for bridges 5/8 min)	–	R = 0 f = 1/4 min r = 1/4 α = 20°	±0 –0 +1/4,–0 +10°,–0°	+3/16,–0‡ ±1/16 ±1/16 +10°,–5°	Flat	S	E, L, Mp

Note A: Not prequalified for gas metal arc welding using short circuiting transfer. Refer to Appendix D.

Note E: Minimum effective throat (E) as shown in Table 2.10.3; S as specified on drawings.

Note L: Butt and T-joints are not prequalified for bridges.

Note Mp: Double-groove welds may have grooves of unequal depth, provided they conform to the limitations of Note E. Also, the effective throat (E), less any reduction, applies individually to each groove.

‡ Fit-up tolerance, SAW: see 3.3.2; for rolled shapes R may be 5/16 inches in thick plates if backing is provided.

Fig. 2.10.1 (continued)—Prequalified partial joint penetration groove welded joints

Single-J-groove weld (B)
Butt joint (B)
T-joint (T)
Corner joint (C)

Welding process	Joint designation	Base metal thickness (U = unlimited)		Groove preparation			Permitted welding positions	Effective throat (E)	Notes
		T_1	T_2	Root opening / Root face / Groove radius / Groove angle	Tolerances As detailed (see 2.10.2)	As fit up (see 3.3.4)			
SMAW	BTC-P8	1/4 min (for bridges 5/16 min)	U	$\alpha = 45°$	+10°,−0°	+10°,−5°	All	S	E, L, V
				R = 0 to 1/8 f = 1/8 min r = 3/8	+1/16,−0 −0 +1/4,−0	±1/16 ±1/16 ±1/16			
GMAW FCAW	BTC-P8-GF	1/4 min (for bridges 5/16 min)	U				All	S	A, E, L, V
				$\alpha = 30°$	+10°,−0°	+10°,−5°			
SAW	C-P8-S	3/8 min (for bridges 7/16 min)	U	$\alpha = 20°$	+10°,−0°	+10°,−5°	Flat	S	E, V
				R = 0 f = 1/4 min r = 1/2	±0 −0 +1/4,−0	+3/16,−0 ‡ ±1/16 ±1/16			
SAW	T-P8-S	3/8 min	U				Flat	S	E, L
				$\alpha = 45°$	+10°,−0°	+10°,−5°			

Note A: Not prequalified for gas metal arc welding using short circuiting transfer. Refer to Appendix D.

Note E: Minimum effective throat (E) as shown in Table 2.10.3; S as specified on drawings.

Note L: Butt and T-joints are not prequalified for bridges.

Note V: For corner joints, the outside groove preparation may be in either or both members, provided the basic groove configuration is not changed and adequate edge distance is maintained to support the welding operations without excessive edge melting.

‡ Fit-up tolerance, SAW: see 3.3.2; for rolled shapes R may be 5/16 inches in thick plates if backing is provided.

Fig. 2.10.1 (continued)—Prequalified partial joint penetration groove welded joints

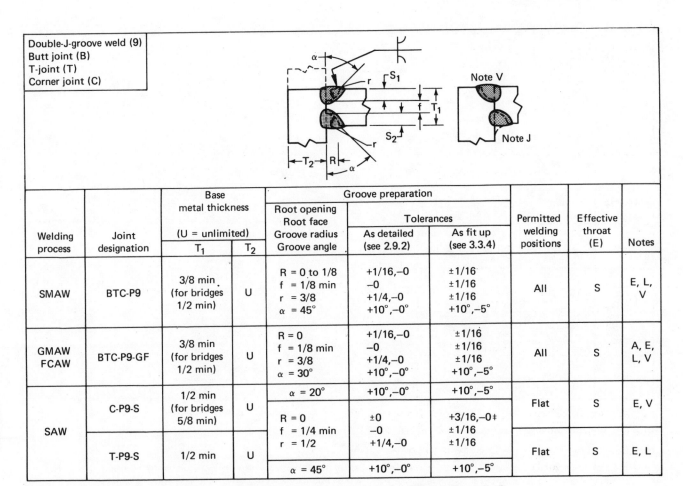

Welding process	Joint designation	Base metal thickness (U = unlimited)		Groove preparation			Permitted welding positions	Effective throat (E)	Notes
				Root opening Root face Groove radius Groove angle	Tolerances				
		T_1	T_2		As detailed (see 2.9.2)	As fit up (see 3.3.4)			
SMAW	BTC-P9	3/8 min (for bridges 1/2 min)	U	R = 0 to 1/8 f = 1/8 min r = 3/8 α = 45°	+1/16,−0 −0 +1/4,−0 +10°,−0°	±1/16 ±1/16 ±1/16 +10°,−5°	All	S	E, L, V
GMAW FCAW	BTC-P9-GF	3/8 min (for bridges 1/2 min)	U	R = 0 f = 1/8 min r = 3/8 α = 30°	+1/16,−0 −0 +1/4,−0 +10°,−0°	±1/16 ±1/16 ±1/16 +10°,−5°	All	S	A, E, L, V
SAW	C-P9-S	1/2 min (for bridges 5/8 min)	U	α = 20°	+10°,−0°	+10°,−5°	Flat	S	E, V
				R = 0 f = 1/4 min r = 1/2	±0 −0 +1/4,−0	+3/16,−0‡ ±1/16 ±1/16			
	T-P9-S	1/2 min	U				Flat	S	E, L
				α = 45°	+10°,−0°	+10°,−5°			

Note A: Not prequalified for gas metal arc welding using short circuiting transfer. Refer to Appendix D.

Note E: Minimum effective throat (E) as shown in Table 2.10.3; S as specified on drawings.

Note L: Butt and T-joints are not prequalified for bridges.

Note Mp: Double-groove welds may have grooves of unequal depth, provided they conform to the limitations of Note E. Also, the effective throat (E), less any reduction, applies individually to each groove.

Note V: For corner joints, the outside groove preparation may be in either or both members, provided the basic groove configuration is not changed and adequate edge distance is maintained to support the welding operations without excessive edge melting.

‡ Fit-up tolerance, SAW: see 3.3.2; for rolled shapes R may be 5/16 inches in thick plates if backing is provided.

Fig. 2.10.1 (continued)—Prequalified partial joint penetration groove welded joints

3. Workmanship

3.1 General

3.1.1 All applicable paragraphs of this section shall be observed in the production and inspection of welded assemblies and structures produced by any of the processes acceptable under this code.

3.1.2 All welding and oxygen-cutting equipment shall be so designed and manufactured and shall be in such condition as to enable qualified welders, welding operators, and tackers to follow the procedures and attain the results prescribed elsewhere in this code.

3.1.3 Welding shall not be done when the ambient temperature is lower than 0° F (−18° C) (see 4.2) when surfaces are wet or exposed to rain, snow, or high wind, or when welders are exposed to inclement conditions.

3.1.4 The sizes and lengths of welds shall be no less than those specified by design requirements and detail drawings, nor shall they be substantially in excess of those requirements without approval. The location of welds shall not be changed without approval.

3.2 Preparation of Base Metal

3.2.1 Surfaces and edges to be welded shall be smooth, uniform, and free from fins, tears, cracks, and other discontinuities which would adversely affect the quality or strength of the weld. Surfaces to be welded and surfaces adjacent to a weld shall also be free from loose or thick scale, slag, rust, moisture, grease, and other foreign material that would prevent proper welding or produce objectionable fumes. Mill scale that can withstand vigorous wire brushing, a thin rust-inhibitive coating, or antispatter compound may remain with the following exception: for girders, all mill scale shall be removed from the surfaces on which flange-to-web welds are to be made by submerged arc welding or by shielded metal arc welding with low-hydrogen electrodes.

3.2.2 In all oxygen cutting, the cutting flame shall be so adjusted and manipulated as to avoid cutting beyond (inside) the prescribed lines. The roughness of oxygen-cut surfaces shall be no greater than that defined by the American National Standards Institute surface roughness value[5] of 1000 μin. (25 μm) for material up to 4 in. (102 mm) thick and 2000 μin. (50 μm) for material 4 in. to 8 in. (203 mm) thick, with the following exception: the ends of members not subject to calculated stress at the ends shall meet the surface roughness value of 2000 μin. Roughness exceeding these values and occasional notches or gouges no more than 3/16 in. (4.8 mm) deep on otherwise satisfactory surfaces shall be removed by machining or grinding. Cut surfaces and edges shall be left free of slag. Correction of discontinuities shall be faired to the oxygen-cut surfaces with a slope not exceeding one in ten. In oxygen-cut edges, occasional notches or gouges less than 7/16 in. (11.1 mm) deep in material up to 4 in. thick, or less than 5/8 in. (15.9 mm) deep in material over 4 in. thick may, with the approval of the Engineer, be repaired by welding. Other discontinuities in oxygen-cut edges shall not be repaired by welding. Any approved weld repairs shall be made by (1) suitably preparing the discontinuity, (2) welding with low-hydrogen electrodes not exceeding 5/32 in. (4.0 mm) diameter, (3) observing the applicable requirements of this code, and (4) grinding the completed weld smooth and flush (see 3.6.3) with the adjacent surface to produce a workmanlike finish.

3.2.3 Visual Inspection and Repair of Plate Cut Edges[6]

3.2.3.1 In the repair and determination of limits of internal discontinuities visually observed on sheared or oxygen-cut edges and caused by entrapped slag or refractory inclusions, deoxidation products, gas pockets, or blow holes, the amount of metal removed shall be the minimum necessary to remove the discontinuity or to determine that the permissible limit is not exceeded. Plate edges may exist at any angle with respect to the rolling direction. All repairs of discontinuities by welding shall conform to the applicable provisions of this code.

---

5. ANSI B46.1, Surface Texture, in microinches (μin.).

6. The requirements of 3.2.3 may not be adequate in cases of tensile load applied through the thickness of the material.

Table 3.2.3
Limits on acceptability and repair of cut-edge discontinuities of plate

Description of discontinuity	Plate repair required
Any discontinuity 1 in. (25.4 mm) in length or less	None, need not be explored.
Any discontinuity over 1 in. (25.4 mm) in length and 1/8 in. (3.2 mm) maximum depth	None, but the depth should be explored.[1]
Any discontinuity over 1 in. (25.4 mm) in length with depth over 1/8 in. (3.2 mm) but not greater than 1/4 in. (6.4 mm)	Remove, need not weld.
Any discontinuity over 1 in. (25.4 mm) in length with depth over 1/4 in. (6.4 mm) but not greater than 1 in.	Completely remove and weld. Aggregate length of welding shall not exceed 20% of the length of the plate edge being repaired.
Any discontinuity over 1 in. (25.4 mm) in length with depth greater than 1 in.	See 3.2.3.3.

1. A spot check of 10 percent of the discontinuities on the oxygen-cut edge in question should be explored by grinding to determine depth. If the depth of any one of the discontinuities explored exceeds 1/8 in. (3.2 mm), then all of the discontinuities remaining on that edge shall be explored by grinding to determine depth. If none of the discontinuities explored in the 10 percent spot check have a depth exceeding 1/8 in. (3.2 mm), then the remainder of the discontinuities on that edge need not be explored.

3.2.3.2 The limits of acceptability and the repair of visually observed edge discontinuities shall be in accordance with Table 3.2.3, in which the length of discontinuity is the visible long dimension on the cut edge of the plate and the depth is the distance that the discontinuity extends into the plate from the cut edge.

3.2.3.3 For discontinuities over 1 in. (25.4 mm) in length with depth greater than 1 in., discovered by visual inspection of cut edges of plate before welding or during examination of welded joints by radiographic or ultrasonic testing, the following procedures should be observed:

(1) Where discontinuities such as (W), (X), or (Y) in Fig. 3.2.3.3 are observed prior to completing the joint, the size and shape of the discontinuity shall be determined by ultrasonic testing. The area of the discontinuity shall be determined as the area of total loss of back reflection, when tested in accordance with the procedure of ASTM A435.

(2) For acceptance, the area of the discontinuity (or the aggregate area of multiple discontinuities) shall not exceed 4% of the plate area (plate length × plate width) with the following exception: if the length of the discontinuity, or the aggregate width of discontinuities on any transverse section, as measured perpendicular to the plate length, exceeds 20% of the plate width, the 4% plate area shall be reduced by the percentage amount of the width exceeding 20%. (For example, if a discontinuity is 30% of the plate width, the area of discontinuity cannot exceed 3.6% of the plate area.) The discontinuity on the cut edge of the plate shall be gouged out to a depth of 1 in. (25.4 mm) beyond its intersection with the surface by chipping, air carbon arc gouging, or grinding, and blocked off by welding with the shielded metal arc process in layers not exceeding 1/8 in. (3.2 mm) in thickness.

(3) If a discontinuity (Z), not exceeding the allowable area in 3.2.3.3 (2), is discovered after the joint has been completed and is determined to be 1 in. (25.4 mm) or more away from the face of the weld, as measured on the plate surface, no repair of the discontinuity is required. If the discontinuity (Z) is less than 1 in. away from the face of the weld, it shall be gouged out to a distance of 1 in. from the fusion zone of the weld by chipping, air carbon arc gouging, or grinding. It shall then be blocked off by welding with the shielded metal arc process for at least four layers not exceeding 1/8 in. (3.2 mm) in thickness per layer. Submerged arc or other welding processes may be used for the remaining layers.

(4) If the area of the discontinuity (W), (X), (Y), or (Z) exceeds the allowable in 3.2.3.3 (2), the plate or subcomponent shall be rejected and replaced, or repaired at the discretion of the Engineer.

(5) The aggregate length of weld repair shall not exceed 20% of the length of the plate edge without approval of the Engineer.

(6) All repairs shall be in accordance with this code. Gouging of the discontinuity may be done from either plate surface or edge.

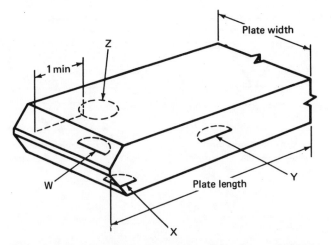

Fig. 3.2.3.3—Edge discontinuities in cut plate

3.2.4 Reentrant corners, except for the corners of weld access cope holes adjacent to a flange, shall be filleted to a radius of no less than 1/2 in. (12.7 mm) for buildings and tubular structures and 3/4 in. (19.0 mm) for bridges. The fillet and its adjacent cuts shall meet without offset or cutting past the point of tangency.

3.2.5 Machining, air carbon arc cutting, oxygen cutting, oxygen gouging, chipping, or grinding may be used for joint preparation, back gouging, or the removal of unacceptable work or metal, except that oxygen gouging shall not be used on steels that are ordered as quenched and tempered or normalized.

3.2.6 Edges of built-up beam and girder webs shall be cut to the prescribed camber with suitable allowance for shrinkage due to cutting and welding. However, moderate variation from the specified camber tolerance may be corrected by a carefully supervised application of heat.

3.2.7 Corrections of errors in camber of quenched and tempered steel must be given prior approval by the Engineer.

3.3 Assembly

3.3.1 The parts to be joined by fillet welds shall be brought into as close contact as practicable. The gap between parts shall not exceed 3/16 in. (4.8 mm) except in cases involving shapes or plates 3 in. (76.2 mm) or greater in thickness if, after straightening and in assembly, the gap cannot be closed sufficiently to meet this tolerance. In such cases, a maximum gap of 5/16 in. (8.0 mm) is acceptable provided a sealing weld or suitable backing material[7] is used to prevent melting-thru. If the separation is 1/16 in. (1.6 mm) or greater, the leg of the fillet weld shall be increased by the amount of the separation

7. Backing to prevent melting-thru may be of flux, glass tape, iron powder, or similar materials; by means of shielded metal arc welding root passes deposited with low-hydrogen electrodes, or other arc welding processes.

or the contractor shall demonstrate that the required effective throat has been obtained.

The separation between faying surfaces of lap joints and of butt welds landing on a backing shall not exceed 1/16 in. (1.6 mm). The use of fillers is prohibited except as specified on the drawings or as specially approved by the Engineer and made in accordance with 2.4.

3.3.2 The parts to be joined by partial joint penetration groove welds parallel to the length of the member, bearing joints excepted, shall be brought into as close contact as practicable. The gap between parts shall not exceed 3/16 in. (4.8 mm) except in cases involving rolled shapes or plates 3 in. (76.2 mm) or greater in thickness if, after straightening and in assembly, the gap cannot be closed sufficiently to meet this tolerance. In such cases a maximum gap of 5/16 in. (8.0 mm) is applicable provided a sealing weld or suitable backing material[8] is used to prevent melting thru and the final weld meets the requirements for effective throat. Tolerances for bearing joints shall be in accordance with the applicable contract specifications.

3.3.3 Abutting parts to be joined by butt welds shall be carefully aligned. Where the parts are effectively restrained against bending due to eccentricity in alignment, an offset not exceeding 10% of the thickness of the thinner part joined, but in no case more than 1/8 in. (3.2 mm), may be permitted as a departure from the theoretical alignment. In correcting misalignment in such cases, the parts shall not be drawn in to a greater slope than 1/2 in. (12.7 mm) in 12 in. (304 mm). Measurement of offset shall be based upon the center line of parts unless otherwise shown on the drawing.

3.3.4 With the exclusion of electroslag and electrogas welding, and with the exception of 3.3.4.1 for root openings in excess of those permitted in the table below and illustrated in Fig. 3.3.4, the dimensions of the cross section of the groove-welded joints which vary from those shown on the detail drawings by more than the following tolerances shall be referred to the Engineer for approval or correction.

	Root not gouged*		Root gouged	
	in.	mm	in.	mm
(1) Root face of joint	±1/16	1.6	Not limited	
(2) Root opening of joints without steel backing	±1/16	1.6	±1/16 −1/8	1.6 3.2
Root opening of joints with steel backing	+1/4 −1/16	6.4 1.6	Not applicable	
(3) Groove angle of joint	+10 deg − 5 deg		+10 deg − 5 deg	

*See 10.13.1.1 (3) for tolerances for complete joint penetration tubular groove welds made from one side without backing.

8. ANSI B46.1, Surface Texture, in microinches (μin.).

3.3.4.1 Root openings wider than those permitted in 3.3.4, but not greater than twice the thickness of the thinner part or 3/4 in. (19 mm), whichever is less, may be corrected by welding to acceptable dimensions prior to joining the parts by welding. Root openings larger than the above may be corrected by welding only with the approval of the Engineer.

3.3.5 Grooves produced by gouging shall be in accordance with groove profile dimensions as specified in Figs. 2.9.1 and 2.10.1.

3.3.6 Members to be welded shall be brought into correct alignment and held in position by bolts, clamps, wedges, guy lines, struts, and other suitable devices, or by tack welds until welding has been completed. The use of jigs and fixtures is recommended where practicable. Suitable allowances shall be made for warpage and shrinkage.

3.3.7 Tack Welds

3.3.7.1 Tack welds shall be subject to the same quality requirements as the final welds except that

(1) Preheat is not mandatory for single pass tack welds which are remelted and incorporated into continuous submerged arc welds.

(2) Discontinuities such as undercut, unfilled craters, and porosity need not be removed before the final submerged arc welding.

3.3.7.2 Tack welds which are incorporated into the final weld shall be made with electrodes meeting the requirements of the final welds and shall be cleaned thoroughly. Multiple pass tack welds shall have cascaded ends.

3.3.7.3 Tack welds not incorporated into final welds shall be removed, except that for buildings they need not be removed unless required by the Engineer.

3.4 Control of Distortion and Shrinkage

3.4.1 In assembling and joining parts of a structure or of built-up members and in welding reinforcing parts to members, the procedure and sequence shall be such as will minimize distortion and shrinkage.

3.4.2 Insofar as practicable, all welds shall be deposited in a sequence that will balance the applied heat of welding while the welding progresses.

3.4.3 The contractor shall prepare a welding sequence for a member or structure which, in conjunction with the joint welding procedures and overall fabrication methods, will produce members or structures meeting the quality requirements specified. The welding sequence and distortion control program shall be submitted to the Engineer, for information and comment, before the start of welding on a member or structure in which shrinkage or distortion is likely to affect the adequacy of the member or structure.

3.4.4 The direction of the general progression in welding on a member shall be from points where the parts are

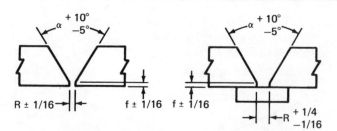

(A) Groove weld without backing — root not gouged

(B) Groove weld with backing — root not gouged

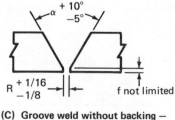

(C) Groove weld without backing — root gouged

Notes:

1. α - groove angle.
2. R - root opening.
3. f - root face.
4. The groove configurations shown are for illustration only.

Fig. 3.3.4—Workmanship tolerances in assembly of groove welded butt joints

relatively fixed in position with respect to each other toward points where they have a greater relative freedom of movement.

3.4.5 Joints expected to have significant shrinkage should usually be welded before joints expected to have lesser shrinkage. They should also be welded with as little restraint as possible.

3.4.6 All shop splices in each component part of a cover-plated beam or built-up member shall be made before the component part is welded to other component parts of the member. Long girders or girder sections may be made by shop-splicing subsections, each made in accordance with this paragraph.

3.4.7 In making welds under conditions of severe external shrinkage restraint, the welding shall be carried continuously to completion or to a point that will insure freedom from cracking before the joint is allowed to cool below the minimum specified preheat and interpass temperature.

3.5 Dimensional Tolerances

3.5.1 The dimensions of welded structural members shall conform to the tolerances of (1) the general specifications governing the work, (2) the special dimensional tolerances in 3.5.1.1 to 3.5.1.12.

3.5.1.1 Permissible variations in straightness of welded columns and primary truss members, regardless of cross section, shall not exceed

Lengths of 45 ft and under:

$$0/8 \text{ in.} \times \frac{\text{No. of ft of total length}}{10}$$

but not over 3/8 in.

Lengths over 45 ft:

$$3/8 \text{ in} + 1/8 \text{ in.} \times \frac{\text{No. of ft of total length} - 45}{10}$$

3.5.1.2 Permissible variations in straightness of welded beams or girders, regardless of cross section, where there is no specified camber or sweep, shall not exceed

$$1/8 \text{ in.} \times \frac{\text{No. of ft of total length}}{10}$$

3.5.1.3 Permissible variation in specified camber of welded beams or girders, regardless of cross section, shall not exceed

−0,
+1/4 in. or

$$-0, +1/4 \text{ in.} \times \frac{\text{No. of ft of test length}}{10}$$

but not to exceed 3/4 in., or

$$-0, +1/8 \text{ in.} \times \frac{\text{No. of ft from nearest end}}{10}$$

whichever is greatest, except that for members whose top flange is embedded in concrete without a designed concrete haunch, the permissible variation, in inches, shall not exceed

$$\pm \frac{\text{No. of ft of total length,}}{160} \text{ or } 1/4 \text{ in.,}$$

whichever is greater.
Note: 3.5.1.3 applies to fabricated pieces before erection.

3.5.1.4 Permissible variation in specified sweep for horizontally curved welded beams or girders is

$$\pm 1/8 \text{ in.} \times \frac{\text{No. of ft of total length}}{10}$$

provided the member has sufficient lateral flexibility to permit the attachment of diaphragms, cross-frames, lateral bracing, etc., without damaging the structural member or its attachments.

3.5.1.5 Permissible lateral variation between the center line of the web and the center line of the flange of built-up H or I members at contact surface shall not exceed 1/4 in.

3.5.1.6 For permissible variations from flatness of web for girders in building and bridge construction, see 8.13 and 9.23, respectively.

3.5.1.7 Combined warpage and tilt of flange of welded beams or girders shall be determined by measuring the offset at the toe of the flange from a line normal to the plane of the web through the intersection of the center line of the web with the outside surface of the flange plate.

This offset shall not exceed 1/100 of the total flange width or 1/4 in. (6.4 mm), whichever is greater, except that abutting parts to be joined by butt welds shall fulfill the requirements of 3.3.3.

3.5.1.8 Permissible Variation from Specified Depth. The maximum permissible variation from specified depth for welded beams and girders, measured at the web center line, shall not exceed

For depths up to 36 in. (0.9 m) incl. . . . . ± 1/8 in. (3.2 mm)
For depths over 36 in. to 72 in. (1.8 m) incl. . . . . . ±3/16 in. (4.8 mm)
For depths over 72 in. . . . . +5/16 in. (8.0 mm)
−3/16 in. (4.8 mm)

3.5.1.9 Bearing at Points of Loading. The bearing ends of bearing stiffeners shall be flush and square with the web and shall have at least 75% of this area in contact with the inner surface of the flanges. The outer surface of the flanges when bearing against a steel base or seat shall fit within 0.010 in. (0.25 mm) for 75% of the projected area of web and stiffeners and not more than 1/32 in. (0.8 mm) for the remaining 25% of the projected area. Girders without stiffeners shall bear on the projected area of the web on the outer flange surface within 0.010 in. and the included angle between web and flange shall not exceed 90 deg in the bearing length.

3.5.1.10 Fit of Intermediate Stiffeners. Where tight fit of intermediate stiffeners is specified, it shall be defined as allowing a gap of up to 1/16 in. (1.6 mm) between stiffener and flange.

3.5.1.11 Straightness of Intermediate Stiffeners. The out-of-straightness variation of intermediate stiffeners shall not exceed 1/2 in. (12.7 mm) with due regard to any members which frame into them.

3.5.1.12 Straightness and Location of Bearing Stiffeners. The out-of-straightness variation of bearing stiffeners shall not exceed 1/4 in. (6.4 mm) up to 6 ft (1.8 m) or 1/2 in. (12.7 mm) over 6 ft. The actual center line of the stiffener shall lie within the thickness of the stiffener as measured from the theoretical center line location.

3.5.1.13 Other Dimensional Tolerances. Dimensional tolerances not covered by 3.5 shall be individually determined and mutually agreed upon by the contractor and the owner with proper regard for erection requirements.

3.6 Weld Profiles

3.6.1 The faces of fillet welds may be slightly convex, flat, or slightly concave as shown in Fig. 3.6 (A), and (B), with none of the unacceptable profiles shown in Fig. 3.6(C). Except at outside corner joints, the convexity shall not exceed the value of 0.1S plus 0.03 in., where S is the actual size of the fillet weld in inches. See Fig. 3.6(B.)

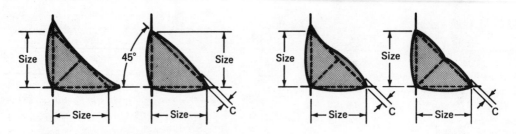

Note: Convexity C shall not exceed 0.1 actual size + 0.03 in.

(A) Desirable fillet weld profiles   **(B) Acceptable fillet weld profiles**

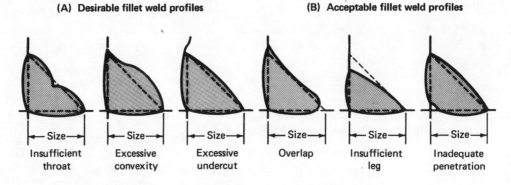

| Insufficient throat | Excessive convexity | Excessive undercut | Overlap | Insufficient leg | Inadequate penetration |

(C) Unacceptable fillet weld profiles

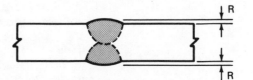

Note: Reinforcement R shall not exceed 1/8 in. See 3.6.2.

(D) Acceptable butt weld profile

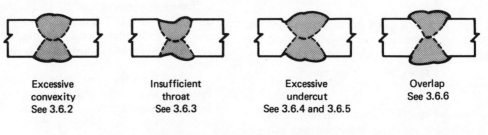

| Excessive convexity See 3.6.2 | Insufficient throat See 3.6.3 | Excessive undercut See 3.6.4 and 3.6.5 | Overlap See 3.6.6 |

(E) Unacceptable butt weld profiles

Fig. 3.6—Acceptable and unacceptable weld profiles

3.6.2 Groove welds shall preferably be made with slight or minimum reinforcement except as may be otherwise provided. In the case of butt and corner joints, the reinforcement shall not exceed 1/8 in. (3.2 mm) in height and shall have gradual transition to the plane of the base metal surface. See Fig. 3.6(D). They shall be free of the discontinuities shown for butt joints in Fig. 3.6(E).

3.6.3 Surfaces of butt joints required to be flush shall be finished so as not to reduce the thickness of the thinner base metal or weld metal by more than 1/32 in. (0.8 mm) or 5% of the thickness, whichever is smaller, nor leave reinforcement that exceeds 1/32 in. However, all reinforcement must be removed where the weld forms part of a faying or contact surface. Any reinforcement must blend smoothly into the plate surfaces with transition areas free from edge weld undercut. Chipping may be used provided it is followed by grinding. Where surface finishing is required, its roughness value[8] shall not exceed 250 μin. (6.3 μm). Surfaces finished to values of over 125 μin. (3.2 μm) through 250 μin. shall be finished parallel to the direction of primary stress. Surfaces finished to values of 125 μin. or less may be finished in any direction.

3.6.3.1 Ends of butt joints required to be flush shall be finished so as not to reduce the width beyond the detailed width or the actual width furnished, whichever is greater, by more than 1/8 in. (3.2 mm) or so as not to leave reinforcement at each end that exceeds 1/8 in. (3.2 mm). Ends of butt welds shall be faired to adjacent plate or shape edges at a slope not to exceed 1 in 10.

3.6.4 For buildings and tubular structures, undercut shall be no more than 0.01 in. (0.25 mm) deep when its direction is transverse to primary tensile stress in the part that is undercut, nor more than 1/32 in. (0.8 mm) for all other situations.

3.6.5 For bridges, undercut shall be no more than 0.01 in. (0.25 mm) deep when the weld is transverse to the primary stress in the part that is undercut. Undercut shall be no more than 1/32 in. (0.8 mm) deep when the weld is parallel to the primary stress in the part that is undercut.

3.6.6 Welds shall be free from overlap.

3.7 Repairs

3.7.1 The removal of weld metal or portions of the base metal may be done by machining, grinding, chipping, oxygen gouging, or air carbon arc gouging. It shall be done in such a manner that the remaining weld metal or base metal is not nicked or undercut. Oxygen gouging shall not be used in quenched and tempered steel. Unacceptable portions of the weld shall be removed without substantial removal of the base metal. Additional weld metal to compensate for any deficiency in size shall be deposited using an electrode preferably smaller than that used for making the original weld, and preferably not more than 5/32 in. (4.0 mm) in diameter. The surfaces shall be cleaned thoroughly before welding.

3.7.2 The contractor has the option of either repairing an unacceptable weld, or removing and replacing the entire weld, except as modified by 3.7.4. The repaired or replaced weld shall be retested by the method originally used, and the same technique and quality acceptance criteria shall be applied. If the contractor elects to repair the weld, it shall be corrected as follows:

3.7.2.1 Overlap or Excessive Convexity. Remove excess weld metal.

3.7.2.2 Excessive Concavity of Weld or Crater, Undersize Welds, Undercutting. Prepare surfaces (see 3.11) and deposit additional weld metal.

3.7.2.3 Excessive Weld Porosity, Excessive Slag Inclusions, Incomplete Fusion. Remove unacceptable portions (see 3.7.1) and reweld.

3.7.2.4 Cracks in Weld or Base Metal. Ascertain the extent of the crack by use of acid etching, magnetic particle inspection, or other equally positive means; remove the crack and sound metal 2 in. (50.8 mm) beyond each end of the crack, and reweld.

3.7.3 Members distorted by welding shall be straightened by mechanical means or by carefully supervised application of a limited amount of localized heat. The temperature of heated areas as measured by approved methods shall not exceed 1100° F (590° C) for quenched and tempered steel nor 1200° F (650° C) (a dull red color) for other steels. The part to be heated for straightening shall be substantially free of stress and from external forces, except those stresses resulting from the mechanical straightening method used in conjunction with the application of heat.

3.7.4 Prior approval of the Engineer shall be obtained for repairs to base metal (other than those required by 3.2), repair of major or delayed cracks, repairs to electroslag and electrogas welds with internal defects, or for a revised design to compensate for deficiencies.

3.7.5 The Engineer shall be notified before improperly fitted and welded members are cut apart.

3.7.6 If, after an unacceptable weld has been made, work is performed which has rendered that weld inaccessible, or has created new conditions that make correction of the unacceptable weld dangerous or ineffectual, then the original conditions shall be restored by removing welds or members or both before the corrections are made. If this is not done, the deficiency shall be compensated for by additional work performed according to an approved revised design.

3.8 Peening

Peening may be used on intermediate weld layers for control of shrinkage stresses in thick welds to prevent cracking. No peening shall be done on the root or surface layer of the weld or the base metal at the edges of the weld. Care should be taken to prevent overlapping or cracking of the weld or base metal.

3.9 Caulking

Caulking of welds shall not be permitted.

---

8. ANSI B46.1, Surface Texture, in microinches (μin.).

3.10 Arc Strikes

Arc strikes outside the area of permanent welds should be avoided on any base metal. Cracks or blemishes caused by arc strikes should be ground to a smooth contour and checked to insure soundness.

3.11 Weld Cleaning

3.11.1 In-process Cleaning. Before welding over previously deposited metal, all slag shall be removed and the weld and adjacent base metal shall be brushed clean. This requirement shall apply not only to successive layers but also to successive beads and to the crater area when welding is resumed after any interruption. It shall not, however, restrict the welding of plug and slot welds in accordance with 4.28 and 4.29.

3.11.2 Cleaning of Completed Welds. Slag shall be removed from all completed welds and the weld and adjacent base metal shall be brushed clean. Welded joints shall not be painted until after the work has been completed and accepted.

3.12 Weld Termination

3.12.1 Welds shall be terminated at the end of a joint in a manner that will insure sound welds. Whenever necessary, this shall be done by use of extension bars and runoff plates.

3.12.2 In building construction, extension bars or runoff plates need not be removed unless required by the Engineer.

3.12.3 In bridge construction, extension bars and runoff plates shall be removed upon completion and cooling of the weld, and the ends of the weld shall be made smooth and flush with the edges of abutting parts.

3.13 Groove Weld Backing

3.13.1 Groove welds made with the use of steel backing shall have the weld metal thoroughly fused with the backing. On bridge structures, steel backing of welds that are transverse to the direction of computed stress shall be removed and the joints shall be ground or finished smooth. Steel backing of welds that are parallel to the direction of stress or are not subject to computed stress need not be removed, unless so specified by the Engineer.

3.13.2 Steel backing of welds used in buildings or tubular structures need not be removed, unless required by the Engineer.

3.13.3 Steel backing shall be made continuous for the full length of the weld. All necessary joints in the steel backing shall be complete joint penetration butt welds meeting all the workmanship requirements of Section 3 of this code.

4. Technique

Part A
General

4.1 Filler Metal Requirements

4.1.1 The electrode, electrode-flux combination, or grade of weld metal for making complete joint penetration butt welds shall be in accordance with Table 4.1.1.

4.1.2 The electrode, electrode-flux combination, or grade of weld metal for complete joint penetration (other than butt welds) or partial joint penetration groove welds, and for fillet welds may be of a lower strength than required for complete joint penetration butt welds provided the weld metal meets the stress requirements (see 8.4, 9.3 or 10.4, whichever is applicable).

4.1.3 After filler metal has been removed from its original package, it shall be protected or stored so that its characteristics or welding properties are not affected.

4.1.4 For exposed, bare, unpainted applications of ASTM A242 and A588 steel requiring weld metal with atmospheric corrosion resistance and coloring characteristics similar to that of the base metal, the electrode, electrode-flux combination, or grade of weld metal shall be in accordance with Table 4.1.4. In multiple-pass welds, the weld metal may be deposited so that at least two layers on all exposed surfaces and edges are deposited with one of the filler metals listed in Table 4.1.4, provided the underlying layers are deposited with one of the filler metals specified in Table 4.1.1.

4.1.5 For single-pass welding, other than electroslag or electrogas, of exposed, bare, unpainted applications of ASTM A242 and A588 steel requiring weld metal with atmospheric corrosion resistance and coloring characteristics similar to that of the base metal, the following variations from Table 4.1.4 may be made.

4.1.5.1 Shielded Metal Arc Welding. Single-pass fillet welds up to 1/4 in. (6.4 mm) maximum and 1/4 in. groove welds made with a single pass, or a single pass each side may be made by using an E70XX low-hydrogen electrode.

4.1.5.2 Submerged Arc Welding. Single-pass fillet welds up to 5/16 in. (8.0 mm) maximum and groove welds made with a single pass, or a single pass each side may be made using an F7X-EXXX electrode-flux combination.

4.1.5.3 Gas Metal Arc Welding. Single-pass fillet welds up to 5/16 in. (8.0 mm) maximum and groove welds made with a single pass, or a single pass each side may be made using an E70S-X electrode.

4.1.5.4 Flux Cored Arc Welding. Single-pass fillet welds up to 5/16 in. (8.0 mm) maximum and groove welds made with a single pass, or a single pass each side may be made using an E70T-X electrode.

4.1.6 For electroslag and electrogas welding of exposed, bare, unpainted applications of ASTM A242 and A588 steel requiring weld metal with atmospheric corrosion resistance and coloring characteristics similar to that of the base metal, the mechanical properties of the weld metal shall meet the requirements of Table 4.16 and the chemical composition requirements of Table 4.1.4.

4.2 Preheat and Interpass Temperature Requirements

With the exclusion of stud welding (see 4.24.7) and electroslag and electrogas welding (see 4.20.5), the minimum preheat and interpass temperatures shall be in accordance with Table 4.2 for the welding process being used and for the higher strength steel being welded. Welding shall not be done when the ambient temperature is lower than 0° F (−18° C). (Zero° F does not mean the ambient environmental temperature but the temperature in the immediate vicinity of the weld. The ambient environmental temperature may be below 0° F but a heated structure or shelter around the area being welded could maintain the temperature adjacent to the weldment at 0° F or higher.) When the base metal is below the specified minimum temperature, it shall be preheated so that the parts on which weld metal is being deposited are at or above the specified minimum temperature for a radius equal to the thickness of the part

Table 4.1.1
Matching filler metal requirements

Group	Steel specification[1,2]		Minimum yield point		Tensile strength range		Electrode specification[3,4]	Minimum yield point		Tensile strength range	
			ksi	MPa	ksi	MPa		ksi	MPa	ksi	MPa
	ASTM A36[5]		36	250	58-80	345-550					
	ASTM A53	Grade B	35	240	60 min	415 min					
	ASTM A106	Grade B	35	240	60 min	415 min	SMAW				
	ASTM A131	Grades A, B, C, CS, D, E	32	220	58-71	400-490	AWS A5.1 or A5.5				
	ASTM A139	Grade B	35	240	60 min	415 min	E60XX or	50	345	67 min	460
	ASTM A381	Grade Y35	35	240	60 min	415 min	E70XX	57	395	70 min	485
	ASTM A500	Grade A	33/39	230/270	45 min	310 min					
		Grade B	42/46	290/320	58 min	400 min					
	ASTM A501		36	250	58 min	400 min	SAW				
	ASTM A516	Grade 55	30	205	55-65	380-450	AWS A5.17 or A5.23				
		Grade 60	32	220	60-72	415-495	F6X-EXXX or	50	345	62-80	425-550
I	ASTM A524	Grade I	35	240	60-85	415-585	F7X-EXXX	60	415	70-95	485-655
		Grade II	30	205	55-80	380-550					
	ASTM A529		42	290	60-85	415-495	GMAW				
	ASTM A570	Grade A	25	170	45 min	310 min	AWS A5.18				
		Grade B	30	210	49 min	340 min	E70S-X or	60	415	72 min	495
		Grade C	33	230	52 min	360 min	E70U-1	60	415	72 min	495
		Grade D	40	275	55 min	380 min					
		Grade E	42	290	58 min	400 min					
	ASTM A573	Grade 65	35	240	65-77	450-530	FCAW				
	ASTM A709	Grade 36[5]	36	250	58-80	345-550	AWS A5.20				
	API 5L	Grade B	35	240	60	415	E60T-X	50	345	62 min	425
	API 5LX	Grade 42	42	290	60	415	E70T-X	60	415	72 min	495
	ABS	Grades A, B, D, CS, DS			58-71	400-490	(Except -2 & -3)				
		Grade E[6]			58-71	400-490					
	ASTM A131	Grades AH32, DH32, EH32	45.5	315	68-85	470-585	SMAW				
II		Grades AH36, DH36, EH36	51	350	71-90	490-620	AWS A5.1 or A5.5				
	ASTM A242[6]		42-50	290-345	63-70 min	435-485	E7015, E7016	57	395	70 min	485
	ASTM A441		42-50	290-345	63-70 min	450-530	E7018, E7028				
	ASTM A516	Grade 65	35	240	65-77	450-530					
		Grade 70	38	260	70-85	485-585					
	ASTM A537	Class 1	50	345	70-90	485-620					

Base Metal

Group	Specification	Grade				
	ASTM A572	Grade 42	42	290	60 min	415 min
		Grade 45	45	310	60 min	415 min
		Grade 50	50	345	65 min	450 min
		Grade 55	55	380	70 min	485 min
	ASTM A588[6]	(4 in. and under)	50	345	70 min	485 min
	ASTM A595	Grade A	55	380	65 min	450 min
		Grades B and C	60	415	70 min	485 min
	ASTM A606[6]		45	310	65 min	450 min
II	ASTM A607	Grade 45	45	310	60 min	415 min
		Grade 50	50	345	65 min	450 min
		Grade 55	55	380	70 min	485 min
	ASTM A618		50	345	70 min	485 min
	ASTM A633	Grades A, B[6]	42	290	63-83	435-570
		Grades C, D (2-1/2 in. and under)	50	345	70-90	485-620
	ASTM A709	Grade 50	50	345	65 min	450 min
		Grade 50W	50	345	70 min	485 min
	API 2H[6]		42	290	62-80	430-550
	ABS	Grades AH32, DH32, EH32	45.5	315	71-90	560-620
		Grades AH36, DH36, EH36[6]	51	350	71-90	560-620
	ASTM A572	Grade 60	60	415	75 min	515 min
		Grade 65	65	450	80 min	550 min
III	ASTM A537	Class 2[6]	60	415	80-110	550-690
	ASTM A633	Grade E[6]	60	415	80-100	550-690

Weld Metal Requirements

Group	Process / Electrode				
II	SAW AWS A5.17 or A5.23 F7X-EXXX	60	415	70-95	485-655
	GMAW AWS A5.18 E70S-X or	60	415	72 min	495
	E70U-1	60	415	72 min	495
	FCAW AWS A5.20 E70T-X (Except -2 & -3)	60	415	72 min	495
III	SMAW AWS A5.5 E8015, E8016 E8018	67	460	80 min	550
	SAW AWS A5.23 F8X-EXXX[7]	68	470	80-100	550-690
	GMAW Grade E80S[7]	65	450	80 min	550
	FCAW Grade E80T[7]	68	470	80-95	550-655

1. In joints involving base metals of two different yield points or strengths, filler metal electrodes applicable to the lower strength base metal may be used, except that if the higher strength base metal requires low hydrogen electrodes, they shall be used.
2. Match API Standard 2B (fabricated tubes) according to steel used.
3. When welds are to be stress-relieved, the deposited weld metal shall not exceed 0.05 percent vanadium.
4. See 4.20 for electrogas and electroslag weld metal requirements.
5. Only low hydrogen electrodes shall be used when welding A36 or A709 Grade 36 steel more than 1 in. thick for bridges.
6. Special welding materials and procedures (e.g., E80XX low alloy electrodes) may be required to match the notch toughness of base metal (for applications involving impact loading or low temperature), or for atmospheric corrosion and weathering characteristics (see 4.1.4).
7. Deposited weld metal shall have a minimum impact strength of 20 ft-lbs (27.1J) at 0° F (−18° C) when Charpy V-notch specimens are used. This requirement is applicable only to bridges.

Table 4.1.1 is continued on p. 46

Table 4.1.1 (continued)
Matching filler metal requirements

Steel specification requirements

Group	Steel specification[1,2]	Minimum yield point ksi	MPa	Tensile strength range ksi	MPa
IV	ASTM A514 Over 2-1/2 in. (63 mm) Grades 100, 100W	90	620	100-130	690-895
	ASTM A709 2-1/2 to 4 in. (63 to 102 mm)	90	620	100-130	690-895
V	ASTM A514 2-1/2 in. (63 mm) and under	100	690	110-130	760-895
	ASTM A517 Grades 100, 100W	100	690	115-135	795-930
	ASTM A709 2-1/2 in. (63 mm) and under	100	690	110-130	760-895

Filler metal requirements

Group	Electrode specification[3,4]	Minimum yield point ksi	MPa	Tensile strength range ksi	MPa
IV	SMAW AWS A5.5 E10015, E10016 E10018	87	600	100 min	690
	SAW AWS A5.23 F10X-EXXX[7]	88	605	100-130	690-895
	GMAW Grade E100S[7]	90	620	100 min	690
	FCAW Grade E100T[7]	88	605	100-115	690-790
V	SMAW AWS A5.5 E11015, E11016 E10018	97	670	110 min	760
	SAW AWS A5.23 F11X-EXXX[7]	98	675	110-130	760-895
	GMAW Grade E110S[7]	98	675	110 min	760
	FCAW Grade E110T[7]	98	675	110-125	760-860

See p. 45 for Notes 1 through 7.

Table 4.1.4
Filler metal requirements for exposed bare applications of ASTM A242 and A588 steel

	Welding process	
Shielded metal arc	Submerged arc	Gas metal arc or flux cored arc[2,4]
AWS A5.5	AWS A5.23	
E8016 or 18-G[1,2]	F7X-EXXX-W[2,3]	
E8016 or 18-B1[2]	F7X-EXXX-B1[2,3]	62 ksi min YP
E8016 or 18-B2[2]	F7X-EXXX-B2[2,3]	(430 MPa)
E8015 or 18-B2L[2]		72 ksi min TS
E8016 or 18-C1	F7X-EXXX-Ni1[3]	(495 MPa)
E8016 or 18-C2	F7X-EXXX-Ni2[3]	Elon. 18% min
E8016 or 18-C3	F7X-EXXX-Ni3[3]	

1. Deposited weld metal shall have the following chemical composition: C, max %, 0.12; Mn, %, 0.50/1.30; P, max %, 0.03; S, max %, 0.04; Si, %, 0.35/0.80; Cu, %, 0.30/0.75; Ni, %, 0.40/0.80; Cr, %, 0.45/0.70.
2. Deposited weld metal shall have a minimum impact strength of Charpy V-notch 20 ft • lb (27.1 J) at 0° F (−18° C) (only applied to bridges).
3. The use of the same type of filler metal having next higher mechanical properties as listed in AWS specifications is permitted.
4. Deposited weld metal shall have a chemical composition the same as that for any one of the weld metals in this table for the shielded metal arc welding process.

Table 4.4.2
Minimum holding time

1/4 in. (6.4 mm) or less	Over 1/4 in. (6.4 mm) through 2 in. (51 mm)	Over 2 in. (51 mm)
15 min	1 hr/in	2 hrs plus 15 min for each additional in. over 2 in. (51 mm)

Table 4.4.3
Alternative stress-relief heat treatment

Decrease in temperature below minimum specified temperature		Minimum holding time at decreased temperature, hours per inch of thickness
°F	°C	
50	10	2
100	38	3
150	66	5
200	93	10

being welded, but not less than 3 in. (76.2 mm) in all directions from the point of welding. Preheat and interpass temperatures must be sufficient to prevent crack formation, and temperatures above the specified minimum may be required for highly restrained welds. In joints involving combinations of base metals, preheat shall be as specified for the higher strength steel being welded.

4.3 Heat Input Control for Quenched and Tempered Steel

When quenched and tempered steels are welded, the heat input shall be restricted in conjunction with the maximum preheat and interpass temperatures required (by reason of base metal thicknesses). The above limitations shall be in strict accordance with the steel producer's recommendations. The use of stringer beads to avoid overheating is strongly recommended. Oxygen gouging of quenched and tempered steels is not permitted.

4.4 Stress Relief Heat Treatment[9]

4.4.1 Where required by the contract drawings or specifications, welded assemblies shall be stress-relieved by heat treating. Finish machining shall preferably be done after stress relieving.

4.4.2 The stress relief treatment shall conform to the following requirements:

(1) The temperature of the furnace shall not exceed 600° F (315° C) at the time the welded assembly is placed in it.

(2) Above 600° F (315° C), the rate of heating[10] shall not be more than 400° F (220° C) per hour divided by the maximum metal thickness of the thicker part in inches,

---

9. Stress relieving of weldments of A514, A517, and A709 Grades 100 and 100W steels is not generally recommended. Stress relieving may be necessary for those applications where weldments must retain dimensional stability during machining or where stress corrosion may be involved, neither condition being unique to weldments involving A514, A517, and A709 Grades 100 and 100W steels. However, the results of notch toughness tests have shown that postweld heat treatment may actually impair weld metal and heat-affected zone toughness, and intergranular cracking may sometimes occur in the grain-coarsened region of the weld heat-affected zone.

10. The rates of heating and cooling need not be less than 100° F (55° C) per hour. However, in all cases, consideration of closed chambers and complex structures may indicate reduced rates of heating and cooling to avoid structural damage due to excessive thermal gradients.

Table 4.2
Minimum preheat and interpass temperature[3,4]

Group I

Steel specification

ASTM A36[2]	
ASTM A53	Grade B
ASTM A106	Grade B
ASTM A131	Grades A, B, C, CS, D, E
ASTM A139	Grade B
ASTM A381	Grade Y35
ASTM A500	Grade A, Grade B
ASTM A501	Grade B
ASTM A516	Grades 55 & 60
ASTM A524	Grades I & II
ASTM A529	
ASTM A570	
ASTM A573	Grade 65
ASTM A709	Grade 36[2]
API 5L	Grade B
API 5LX	Grade 42
ABS	Grades A, B, D, CS, DS; Grade E

Welding process: Shielded metal arc welding with other than low hydrogen electrodes

Thickness of thickest part at point of welding, in.	mm	Minimum temperature, °F	°C
Up to 3/4	19 incl.	None[1]	
Over 3/4 thru 1-1/2	19 – 38 incl.	150	66
Over 1-1/2 thru 2-1/2	38 – 64 incl.	225	107
Over 2-1/2	64	300	150

Group II

Steel specification

ASTM A36[2]	All grades
ASTM A53	Grade B
ASTM A106	Grade B
ASTM A131	Grades A, B, C, CS, D, E; AH 32 & 36; DH 32 & 36; EH 32 & 36
ASTM A139	Grade B
ASTM A242	
ASTM A381	Grade Y35
ASTM A441	
ASTM A500	Grade A, Grade B
ASTM A501	Grade B
ASTM A516	Grades 55 & 60, 65 & 70
ASTM A524	Grades I & II
ASTM A529	
ASTM A537	Classes 1 & 2
ASTM A570	All grades
ASTM A572	Grades 42, 45, 50
ASTM A573	Grade 65
ASTM A588	
ASTM A595	Grades A, B, C
ASTM A606	
ASTM A607	Grades 45, 50, 55
ASTM A618	
ASTM A633	Grades A, B; Grades C, D
ASTM A709	Grades 36, 50, 50W
API 5L	Grade B
API 5LX	Grade 42
API Spec. 2H	
ABS	Grades AH 32 & 36, DH 32 & 36, EH 32 & 36
ABS	Grades A, B, D, CS, DS; Grade E

Welding process: Shielded metal arc welding with low hydrogen electrodes, submerged arc welding, gas metal arc welding, flux cored arc welding

Thickness of thickest part at point of welding, in.	mm	Minimum temperature, °F	°C
Up to 3/4	19 incl.	None[1]	
Over 3/4 thru 1-1/2	19 – 38 incl.	50	10
Over 1-1/2 thru 2-1/2	38 – 64 incl.	150	66
Over 2-1/2	64	225	107

	Steel	Grade	Welding process	Thickness (in.)	Thickness (mm)	°F	°C
III	ASTM A572	Grades 55, 60, 65	Shielded metal arc welding with low hydrogen electrodes, submerged arc welding, gas metal arc welding, flux cored arc welding	Up to 3/4	19 incl.	50	10
	ASTM A633	Grade E		Over 3/4 thru 1-1/2	19–38 incl.	150	66
				Over 1-1/2 thru 2-1/2	38–64 incl.	225	107
				Over 2-1/2	64	300	150
IV	ASTM A514 ASTM A517 ASTM A709	Grades 100 & 100W	Shielded metal arc welding with low hydrogen electrodes, submerged arc welding with carbon or alloy steel wire, neutral flux, gas metal arc welding or flux cored arc welding	Up to 3/4	19 incl.	50	10
				Over 3/4 thru 1-1/2	19–38 incl.	125	50
				Over 1-1/2 thru 2-1/2	38–64 incl.	175	80
				Over 2-1/2	64	225	107
V	ASTM A514 ASTM A709	Grades 100 & 100W	Submerged arc welding with carbon steel wire, alloy flux	Up to 3/4	19 incl.	50	10
				Over 3/4 thru 1-1/2	19–38 incl.	200	95
				Over 1-1/2 thru 2-1/2	38–64 incl.	300	150
				Over 2-1/2	64	400	205

Notes:

A. For modification of preheat requirements for submerged arc welding with multiple electrodes, see 4.15.5.

B. Zero° F (−18° C) does not mean the ambient environmental temperature but the temperature in the immediate vicinity of the weld. The ambient environmental temperature may be below 0° F, but a heated structure or shelter around the area being welded could maintain the temperature adjacent to the weldment at 0° F or higher.

C). When the base metal is below the temperature listed for the welding process being used and the thickness of material being welded, it shall be preheated (except as otherwise provided) in such manner that the surfaces of the parts on which weld metal is being deposited are at or above the specified minimum temperature for a distance equal to the thickness of the part being welded, but not less than 3 in. (76 mm), both laterally and in advance of the welding. Preheat and interpass temperatures must be sufficient to prevent crack formation. Temperature above the minimum shown may be required for highly restrained welds. For A514, A517, and A709 Grades 100 and 100W steel the maximum preheat and interpass temperature shall not exceed 400° F (205° C) for thickness up to 1-1/2 in. (38 mm) inclusive, and 450° F (230° C) for greater thicknesses. Heat input when welding A514, A517, and A709 Grades 100 and 100W steel shall not exceed the steel producer's recommendations.

1. When the base metal temperature is below 32° F (0° C), the base metal shall be preheated to at least 70° F (21° C) and this minimum temperature maintained during welding.

2. Only low hydrogen electrodes shall be used when welding A36 or A709 Grade 36 steel more than 1 in. thick for bridges.

3. Welding shall not be done when the ambient temperature is lower than 0° F (−18°

4. In joints involving combinations of base metals, preheat shall be as specified for the higher strength steel being welded.

but in no case more than 400° F per hour. During the heating period, variation in temperature throughout the portion of the part being heated shall be no greater than 250° F (140° C) within any 15 ft (4.6 m) interval of length.

(3) After a maximum temperature of 1100° F (590° C) is reached on quenched and tempered steels, or a mean temperature range between 1100 and 1200° F (650° C) is reached on other steels, the temperature of the assembly shall be held within the specified limits for a time not less than specified in Table 4.4.2, based on weld thickness. When the specified stress relief is for dimensional stability, the holding time shall be not less than specified in Table 4.4.2, based on the thickness of the thicker part. During the holding period there shall be no difference greater than 150° F (83° C) between the highest and lowest temperature throughout the portion of the assembly being heated.

(4) Above 600° F (315° C), cooling shall be done in a closed furnace or cooling chamber at a rate[10] no greater than 500° F (260° C) per hour divided by the maximum metal thickness of the thicker part in inches, but in no case more than 500° F per hour. From 600° F (315° F), the assembly may be cooled in still air.

4.4.3 Alternatively, when it is impractical to postweld heat treat to the temperature limitations stated in 4.4.2, welded assemblies may be stress-relieved at lower temperatures for longer periods of time, as given in Table 4.4.3.

Part B
Shielded Metal Arc Welding

4.5 Electrodes for Shielded Metal Arc Welding

4.5.1 Electrodes for shielded metal arc welding shall conform to the requirements of the latest edition of AWS A5.1, Specification for Mild Steel Covered Arc-Welding Electrodes, or to the requirements of AWS A5.5, Specification for Low-Alloy Steel Covered Arc-Welding Electrodes.

4.5.2 Low-Hydrogen Electrode Storage Conditions. All electrodes having low-hydrogen coverings conforming to AWS A5.1 shall be purchased in hermetically sealed containers or shall be dried for at least two hours between 450° F (230° C) and 500° F (260° C) before they are used. Electrodes having a low-hydrogen covering conforming to AWS A5.5 shall be purchased in hermetically sealed containers or shall be dried at least one hour at temperatures between 700° F (370° C) and

800° F (430° C) before being used. Electrodes shall be dried prior to use if the hermetically sealed container shows evidence of damage. Immediately after opening of the hermetically sealed container or removal of the electrodes from drying ovens, electrodes shall be stored in ovens held at a temperature of at least 250° F (120° C). After the opening of hermetically sealed containers or removal from drying or storage ovens, electrode exposure to the atmosphere shall not exceed the requirements of either 4.5.2.1 or 4.5.2.2.

4.5.2.1 Approved Atmospheric Exposure Time Periods. After hermetically sealed containers are opened or after electrodes are removed from drying or storage ovens, the electrode exposure to the atmosphere shall not exceed the values shown in Column A, Table 4.5.2, for the specific electrode classification.

4.5.2.2 Alternative Atmosphere Exposure Time Periods Established by Tests. The alternative exposure time values shown in Column B in Table 4.5.2 may be used provided testing by the user establishes the maximum allowable time. The testing shall be performed in accordance with AWS 5.5, Section 25, for each electrode classification and each electrode manufacturer. Such tests shall establish that the maximum moisture content values of AWS A5.5 (Table 7) are not exceeded. Additionally, E70XX (AWS A5.1 or A5.5) low-hydrogen electrode coverings shall be limited to a maximum moisture content not exceeding 0.4% by weight.

These electrodes shall not be used at relative humidity-temperature combinations that exceed either the relative

Table 4.5.2
Permissible atmospheric exposure of low-hydrogen electrodes

Electrode	Column A (hours)	Column B (hours)
A5.1		
E70XX	4 max	Over 4 to 10 max
A5.5		
E70XX	4 max	Over 4 to 10 max
E80XX	2 max	Over 2 to 10 max
E90XX	1 max	Over 1 to 5 max
E100XX	1/2 max	Over 1/2 to 4 max
E110XX	1/2 max	Over 1/2 to 4 max

Notes:

1. Column A: Electrodes exposed to atmosphere for longer periods than shown shall be redried before use.

2. Column B: Electrodes exposed to atmosphere for longer periods than those established by testing shall be redried before use.

humidity or moisture content in the air that prevailed during the testing program.[11]

4.5.3 Electrode Restrictions for A514 or A517 Steels. When used for welding ASTM A514 or A517 steels, electrodes of any classification lower than E100XX shall be dried at least one hour at temperatures between 700 and 800° F (370 and 430° C) before being used, whether furnished in hermetically sealed containers or otherwise.

4.5.4 Redrying Electrodes. Electrodes that conform to the provisions of 4.5.2 shall subsequently be redried no more than one time. Electrodes that have been wet shall not be used.

4.5.5 Manufacturer's Certification. When requested by the Engineer, the contractor or fabricator shall furnish an electrode manufacturer's certification that the electrode will meet the requirements of the classification.

4.6 Procedures for Shielded Metal Arc Welding

4.6.1 The work shall be positioned for flat position welding whenever practicable.

4.6.2 The classification and size of electrode, arc length, voltage, and amperage shall be suited to the thickness of the material, type of groove, welding positions, and other circumstances attending the work. Welding current shall be within the range recommended by the electrode manufacturer.

4.6.3 The maximum diameter of electrodes shall be as follows:

    4.6.3.1 5/16 in. (8.0 mm) for all welds made in the flat position, except root passes.

    4.6.3.2 1/4 in. (6.4 mm) for horizontal fillet welds.

    4.6.3.3 1/4 in. (6.4 mm) for root passes of fillet welds made in the flat position and groove welds made in the flat position with backing and with a root opening of 1/4 in. or more.

    4.6.3.4 5/32 in. (4.0 mm) for welds made with EXX14 and low-hydrogen electrodes in the vertical and overhead positions.

    4.6.3.5 3/16 in. (4.8 mm) for root passes of groove welds and for all other welds not included under 4.6.3.1, 4.6.3.2, 4.6.3.3, and 4.6.3.4.

4.6.4 The minimum size of a root pass shall be sufficient to prevent cracking.

4.6.5 The maximum thickness of root passes in groove welds shall be 1/4 in. (6 mm).

11. For proper application of this provision, see Appendix J for the temperature-moisture content chart and its examples. The chart shown in Appendix J, or any standard psychrometric chart, must be used in the determination of temperature-relative humidity limits.

4.6.6 The maximum size of single pass fillet welds and root passes of multiple pass fillet welds shall be

    4.6.6.1 3/8 in. (9.5 mm) in the flat position

    4.6.6.2 5/16 in. (8.0 mm) in the horizontal or overhead positions

    4.6.6.3 1/2 in. (12.7 mm) in the vertical position

4.6.7 The maximum thickness of layers subsequent to root passes of groove and fillet welds shall be

    4.6.7.1 1/8 in. (3 mm) for subsequent layers of welds made in the flat position

    4.6.7.2 3/16 in. (4 mm) for subsequent layers of welds made in the vertical, overhead, or horizontal positions

4.6.8 The progression for all passes in vertical position welding shall be upward, except that undercut may be repaired vertically downwards when preheat is in accordance with Table 4.2, but not lower than 70° F (21° C). However, when tubular products are welded, the progression of vertical welding may be upwards or downwards but only in the direction or directions for which the welder is qualified.

4.6.9 Complete joint penetration groove welds made without the use of steel backing shall have the root gouged to sound metal before welding is started from the second side, except as permitted by 10.13.

Part C
Submerged Arc Welding

4.7 General Requirements

4.7.1 Submerged arc welding may be performed with one or more single electrodes, one or more parallel electrodes,[12] or combinations of single and parallel electrodes. The spacing between arcs shall be such that the slag cover over the weld metal produced by a leading arc does not cool sufficiently to prevent the proper weld deposit of a following electrode. Submerged arc welding with multiple electrodes may be used for any groove or fillet weld pass.

4.7.2 The following paragraphs (4.7.3-4.7.8) governing the use of submerged arc welding are suitable for any steel included in 8.2, 9.2, or 10.2 other than those of the quenched and tempered group. Concerning the latter group, it is necessary to comply with the steel producer's recommendation for maximum permissible heat input and preheat combinations. Such considerations must include the additional heat input produced in simultaneous welding on the two sides of a common member.

4.7.3 The diameter of electrodes shall not exceed 1/4 in. (6.4 mm).

12. See Appendix I.

4.7.4 Surfaces on which submerged arc welds are to be deposited and adjacent faying surfaces shall be clean and free of moisture as specified in 3.2.1.

4.7.5 When the joint to be welded requires specific root penetration, as in joints B-L1-S, TC-L1-S, B-L2b-S, C-L2b-S, B-U3a-S, B-L3-S, TC-L4-S, TC-U5-S, and B-U7-S (illustrated in Fig. 2.9.1), unless the joint is back-gouged, the contractor shall prepare a sample joint and macroetched cross section to demonstrate that the proposed welding procedure will attain the required root penetration. The Engineer at his discretion may accept a radiograph of a test joint or recorded evidence in lieu of the test specified in this paragraph. (The Engineer should accept properly documented evidence of previous qualification tests.)

4.7.6 Roots of groove or fillet welds may be backed by copper, flux, glass tape, iron powder, or similar materials to prevent melting thru. They may also be sealed by means of root passes deposited with low-hydrogen electrodes if shielded metal arc welding is used, or by other arc welding processes.

4.7.7 Neither the depth nor the maximum width in the cross section of weld metal deposited in each weld pass shall exceed the width at the surface of the weld pass (see Fig. 4.7.7). This requirement may be waived only if the testing of a welding procedure to the satisfaction of the Engineer has demonstrated that such welds exhibit freedom from cracks, and the same welding procedure and flux-electrode classifications are used in construction.

4.7.8 Tack welds (in the form of fillet welds 3/8 in. [9.5 mm] or smaller, or in the roots of joints requiring specific root penetration) shall not produce objectionable changes in the appearance of the weld surface or result in decreased penetration. Tack welds not conforming to the preceding requirements shall be removed or reduced in size by any suitable means before welding. Tack welds in the root of a joint with steel backing less than 5/16 in. (8.0 mm) thick shall be removed or made continuous for the full length of the joint using shielded metal arc welding with low-hydrogen electrodes.

4.8 Electrodes and Fluxes for Submerged Arc Welding

4.8.1 The bare electrodes and flux used in combination for submerged arc welding of steels shall conform to the requirements in the latest edition of AWS A5.17, Specification for Bare Mild Steel Electrodes and Fluxes for Submerged Arc Welding, or to the requirements of the latest edition of AWS A5.23, Specification for Bare Steel Electrodes and Fluxes for Submerged Arc Low-Alloy Steel Weld Metal.

4.8.2 When requested by the Engineer, the contractor or fabricator shall furnish an electrode manufacturer's certification that the electrode and flux combination will meet the requirements of the classification or grade.

4.9 Condition of Flux

Flux used for submerged arc welding shall be dry and free of contamination from dirt, mill scale, or other foreign material. All flux shall be purchased in packages that can be stored, under normal conditions, for at least six months without such storage affecting its welding characteristics or weld properties. Flux from damaged packages shall be discarded or shall be dried at a minimum temperature of 250° F (120° C) for one hour before use. Flux shall be placed in the dispensing system immediately upon opening a package, or, if used from an unopened package, the top one inch shall be discarded. Flux that has been wet shall not be used. Flux fused in welding shall not be reused.

4.10 Procedures for Submerged Arc Welding with a Single Electrode

4.10.1 All submerged arc welds except fillet welds shall be made in the flat position. Fillet welds may be made in either the flat or horizontal position, except that single pass fillet welds made in the horizontal position shall not exceed 5/16 in. (8.0 mm).

4.10.2 The thickness of weld layers, except root and surface layers, shall not exceed 1/4 in. (6.4 mm). When the root opening is 1/2 in. (12.7 mm) or greater, a multiple pass, split-layer technique shall be used. The split-layer technique shall also be used in making multiple pass welds when the width of the layer exceeds 5/8 in. (15.9 mm).

4.10.3 The welding current, arc voltage, and speed of travel shall be such that each pass will have complete fusion with the adjacent base metal and weld metal and there will be no overlap or undue undercutting. The maximum welding current to be used in making a groove weld for any pass that has fusion to both faces of the groove shall be 600 A, except that the final layer may be made using a higher current. The maximum current to be used for making fillet welds in the flat position shall be 1000 A.

4.11 Procedures for Submerged Arc Welding with Multiple Electrodes

4.11.1 Submerged arc welds with multiple electrodes, except fillet welds, shall be made in the flat position. Fillet welds may be made in either the flat or horizontal position, except that single pass multiple electrode fillet

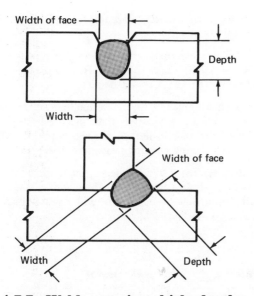

Fig. 4.7.7—Weld pass in which depth and width exceed the width of the weld face

welds made in the horizontal position shall not exceed 1/2 in. (12.7 mm).

4.11.2 The thickness of weld layers is not limited. In making the root pass of a groove weld, single or multiple electrodes may be used. Backing bars or root faces shall be of adequate thickness to prevent melting thru. When the width of a surface in a groove on which a layer of weld metal is to be deposited exceeds 1/2 in. (12.7 mm), multiple electrodes shall be displaced laterally or a split-layer technique used to assure adequate corner fusion. When the width of a previously deposited layer exceeds 1 in. (25.4 mm) and two electrodes only are used, a split-layer technique with electrodes in tandem shall be employed.

4.11.3 The welding current, arc voltage, speed of travel, and relative location of electrodes shall be such that each pass will have complete fusion with the adjacent base metal and weld metal and such that there will be no depressions or undue undercutting at the toe of the weld. Excessive concavity of initial passes shall be avoided to prevent cracking in roots of joints under restraint.

4.11.3.1 The maximum welding current in making a groove weld shall be

(1) 700 A for any single electrode or for parallel electrodes when making the root layer in a groove having no root opening and which does not fill the groove.

(2) 750 A for any single electrode or 900 A for parallel electrodes when making the root pass in a groove having steel backing or a spacer bar.

(3) 1000 A for any single electrode or 1200 A for parallel electrodes for all other passes except the final layer.

(4) For the final layer there is no restriction on welding current.

4.11.3.2 The maximum welding current to be used in making a fillet weld shall be 1000 A for any single electrode or 1200 A for parallel electrodes.

4.11.4 Multiple electrode welds may also be made in the root of groove or fillet welds using gas metal arc welding followed by multiple submerged arcs, provided that the gas metal arc welding conforms to the requirements of Part D of this Section, and provided the spacing between the gas shielded arc and the first following submerged arc does not exceed 15 in. (380 mm).

4.11.5 Preheat and interpass temperatures for multiple-electrode submerged arc welding shall be selected in accordance with Table 4.2. For single pass groove or fillet welds, for combinations of metals being welded and the heat input involved, and with the approval of the Engineer, preheat and interpass temperatures may be established which are sufficient to reduce the hardness in the heat-affected zones of the base metal to less than 225 Vickers hardness number for steel having a minimum specified tensile strength not exceeding 60 000 psi (415 MPa) and 280 Vickers hardness number for steel having a minimum specified tensile strength greater than 60 000 but not exceeding 70 000 psi (485 MPa).[13]

4.11.5.1 Hardness determinations of the heat-affected zones shall be made on

(1) Initial macroetch cross sections of a sample test specimen, and

(2) The surface of the member during the progress of the work. The surface shall be ground prior to hardness testing.

(a) The frequency of such heat-affected zone testing shall be at least one test area per weldment on the thicker metal involved in a joint for each 50 ft (15.2 m) of groove welds or pair of fillet welds.

(b) These hardness determinations may be discontinued after the procedure has been established to the satisfaction of the Engineer.

4.11.5.2 No reduction of the preheat requirements of Table 4.2 will be permitted for fillet welds 3/8 in. (9.6 mm) and under in size.

Part D
Gas Metal Arc and
Flux Cored Arc Welding

4.12 Electrodes

4.12.1 The electrodes and shielding for gas metal arc welding or flux cored arc welding for producing weld

---

13. The Vickers hardness number shall be determined in accordance with ASTM E92. If another method of hardness testing is to be used, the equivalent hardness number shall be determined from ASTM E140, and testing shall be performed according to the applicable ASTM specification.

metal with minimum specified yield strengths of 60 000 psi (415 MPa), or less, shall conform to the requirements of the latest edition of AWS A5.18, Specification for Mild Steel Electrodes for Gas Metal Arc Welding, or AWS A5.20, Specification for Mild Steel Electrodes for Flux Cored Arc Welding, as applicable.

4.12.2 For weld metal having a minimum specified yield strength greater than 60 000 psi (415 MPa), the user shall demonstrate that each combination of electrode and shielding proposed for use will produce low alloy weld metal having the mechanical properties listed in Table 4.12. The mechanical properties shall be determined from a multiple pass weld made in accordance with the test requirements of the latest edition of AWS A5.18 or A5.20, as applicable. When an applicable AWS filler metal spec-

ification is issued, it will control, and testing by the user will not be required.

4.12.3 The mechanical property tests required in 4.12.2 for Grades E100S, E110S, E100T, and E110T shall be made using ASTM A514 base metal.

4.12.4 The Engineer at his discretion may accept recorded evidence of a combination that has been satisfactorily tested by the user in lieu of the test required in 4.12.2, provided the same welding procedure is used.

4.12.5 When requested by the Engineer, the contractor or fabricator shall furnish the electrode manufacturer's certification that the electrode will meet the above requirements of classification or grade.

Table 4.12
Mechanical properties of low-alloy GMAW and FCAW weld metal

U.S. Customary Units	SI Units
GMAW Grade E80S and FCAW Grade E80T	**GMAW Grade E80S and FCAW Grade E80T**
Tensile strength, psi min 80 000	Tensile strength, MPa min 550
Yield strength, psi min 65 000	Yield strength, MPa min 450
Elongation in 2 in., % min. 18	Elongation in 51 mm, % min 18
Impact strength, min	Impact strength, min
Charpy V-notch* . . . . 20 ft-lb at 0° F	Charpy V-notch* . . . . 27.1 J at −18° C
GMAW Grade E90S and FCAW Grade E90T	**GMAW Grade E90S and FCAW Grade E90T**
Tensile strength, psi min 90 000	Tensile strength, MPa min 620
Yield strength, psi min 78 000	Yield strength, MPa min 540
Elongation in 2 in., % min 17	Elongation in 51 mm, % min 17
Impact strength, min	Impact strength, min
Charpy V-notch* . . . . 20 ft-lb at 0° F	Charpy V-notch* . . . . 27.1 J at −18° C
GMAW Grade E100S and FCAW Grade E100T	**GMAW Grade E100S and FCAW Grade E100T**
Tensile strength, psi min 100 000	Tensile strength, MPa min 690
Yield strength, psi min 90 000	Yield strength, MPa min 620
Elongation in 2 in., % min 16	Elongation in 51 mm, % min 16
Impact strength, min	Impact strength, min
Charpy V-notch* . . . . 20 ft-lb at 0° F	Charpy V-notch* . . . . 27.1 J at −18° C
GMAW Grade E110S and FCAW Grade E110T	**GMAW Grade E110S and FCAW Grade E110T**
Tensile strength, psi min 110 000	Tensile strength, MPa min 750
Yield strength, psi min 98 000	Yield strength, MPa min 675
Elongation in 2 in., % min 15	Elongation in 51 mm, % min 15
Impact strength, min	Impact strength, min
Charpy V-notch* . . . . 20 ft-lb at 0° F	Charpy V-notch* . . . . 27.1 J at −18° C

*For bridge application only. Base metal requirements, if more restrictive, shall be met.

4.13 Shielding Gas

A gas or gas mixture used for shielding in gas metal arc welding or flux cored arc welding shall be of a welding grade having a dew point of −40° F (−40° C) or lower. When requested by the Engineer, the contractor or fabricator shall furnish the gas manufacturer's certification that the gas or gas mixture is suitable for the intended application and will meet the dew point requirement.

4.14 Procedures for Gas Metal Arc and Flux Cored Arc Welding with Single Electrode

4.14.1 The following are the requirements for prequalified procedures that are exempt from qualification testing.

4.14.1.1 Electrodes shall be dry and in suitable condition for use.

4.14.1.2 The maximum electrode diameter shall be 5/32 in. (4.0 mm) for the flat and horizontal positions, 3/32 in. (2.4 mm) for the vertical position, and 5/64 in. (2.0 mm) for the overhead position.

4.14.1.3 The maximum size of a fillet weld made in one pass shall be 1/2 in. (12.7 mm) for the flat and vertical positions, 3/8 in. (9.5 mm) for the horizontal position, and 5/16 in. (8.0 mm) for the overhead position.

4.14.1.4 The thickness of weld layers, except root and surface layers, shall not exceed 1/4 in. (6.4 mm). When the root opening of a groove weld is 1/2 in. (12.7 mm) or greater, a multiple pass, split-layer technique shall be used. The split-layer technique shall also be used in making all multiple pass welds when the width of the layer exceeds 5/8 in. (15.9 mm).

4.14.1.5 The welding current, arc voltage, gas flow, mode of metal transfer, and speed of travel shall be such that each pass will have complete fusion with adjacent base metal and weld metal and there will be no overlap or excessive porosity or undercutting.

4.14.1.6 The progressions for all passes of vertical position welding shall be upwards except that undercut may be repaired vertically downwards when preheat is in accordance with Table 4.2 but no lower than 70° F (21° C). In tubular structures the progression of vertical welding may be upwards or downwards, but only in the direction or directions for which the welder is qualified.

4.14.2 Complete joint penetration groove welds made without the use of backing shall have the root of the initial weld gouged, chipped, or otherwise removed to all but traces[14] of the root of the initial weld before welding is started from the second side.

4.14.3 Gas metal arc or flux cored arc welding with external gas shielding shall not be done in a draft or wind

_____
14. Intermittent remnant of root.

unless the weld is protected by a shelter. Such shelter shall be of material and shape appropriate to reduce wind velocity in the vicinity of the weld to a maximum of five miles per hour.

4.14.4 To prevent melting thru, roots of groove or fillet welds may be backed by copper, flux, glass tape, iron powder, or similar materials, or sealed by means of root passes deposited by shielded metal arc welding with low-hydrogen electrodes, or other arc welding processes.

Part E
Electroslag and Electrogas Welding

4.15 Qualification of Process, Procedures, and Joint Details

4.15.1 Prior to use, the contractor shall prepare a procedure specification and qualify each procedure for each process to be used according to the requirements in Section 5. The procedure specification shall include the joint details, filler metal type and diameter, amperage, voltage (type and polarity), speed of vertical travel if not an automatic function of arc length or deposition rate, oscillation (traverse speed, length, and dwell time), type of shielding including flow rate and dew point of gas or type of flux, type of molding shoe, postweld heat treatment if used, and other pertinent information.

4.15.2 Electroslag or electrogas welding of steels that are ordered as quenched and tempered or normalized is not permitted.

4.15.3 When required by contract drawings or specifications, impact tests shall be included in the welding procedure qualification. The impact tests, requirements, and procedure shall be in accordance with the provisions of Appendix C.

4.15.4 The Engineer, at his discretion, may accept evidence of previous qualification of the joint welding procedures to be employed.[15]

4.16 All-Weld-Metal Tension Test Requirements

Prior to use, the contractor shall demonstrate by the tests prescribed in Part B of Section 5 that each combination of shielding and filler metal will produce welds having

_____
15. The Engineer should accept properly documented evidence of previous qualification tests.

the mechanical properties listed in Table 4.16 when welded in accordance with the procedure specification. The Engineer at his discretion may accept recorded evidence of a combination that has been satisfactorily tested in lieu of the testing required provided the same welding procedure is used.[15]

4.17 Condition of Electrodes and Guide Tubes

Electrodes and consumable guide tubes shall be dry, clean, and in suitable condition for use.

4.18 Shielding Gas

A gas or a gas mixture used for shielding for electrogas welding shall be of a welding grade and have a dew point of −40° F (−40° C) or lower. When requested by the Engineer, the contractor or fabricator shall furnish the gas manufacturer's certification that the gas or gas mixture is suitable for the intended application and will meet the dew point requirements.

4.19 Condition of Flux

Flux used for electroslag welding shall be dry and free of contamination from dirt, mill scale, or other foreign material. All flux shall be purchased in packages that can be stored, under normal conditions, for at least six months without such storage affecting its welding characteristics or weld properties. Flux from packages damaged in transit or in handling shall be discarded or shall be dried at a minimum temperature of 250° F (120° C) for one hour before use. Flux that has been wet shall not be used.

4.20 Procedures for Electroslag and Electrogas Welding

4.20.1 Gas to be used for shielding shall be of a welding grade and shall meet all requirements of the procedure specification. When mixed at the welding site, suitable meters shall be used for proportioning the gases. Percentage of gases shall conform to the requirements of the procedure specification.

4.20.2 Electrogas welding shall not be done in a draft or wind of a velocity greater than five miles per hour unless the weld is protected by a shelter. This shelter shall be of a material and shape appropriate to reduce wind velocity in the vicinity of the weld surface to a maximum of five miles per hour.

4.20.3 The type and diameter of the electrodes used shall meet the requirements of the procedure specification.

4.20.4 Welds shall be started in such a manner as to permit sufficient heat buildup for complete fusion of the weld metal to the groove faces of the joint. Welds which have been stopped at any point in the weld joint for a sufficient amount of time for the slag or weld pool to begin to solidify may be restarted and completed, provided the completed weld is examined by ultrasonic testing for a minimum of 6 in. (150 mm) on either side of the restart and, unless prohibited by joint geometry, also confirmed by radiographic testing. All such restart locations shall be recorded and reported to the Engineer.

4.20.5 Because of the high heat input characteristic of these processes, preheating is not normally required. However, no welding shall be performed when the temperature of the base metal at the point of welding is below 32° F (0° C).

4.20.6 Welds having defects prohibited by 8.15 or 9.25 shall be repaired as permitted by 3.7 utilizing a qualified welding process, or the entire weld shall be removed and replaced.

Part F
Stud Welding

4.21 Scope

Part F contains provisions for the installation and inspection of steel studs welded to steel, to attach members and connection devices to concrete (as concrete anchors and as shear connectors in composite steel-concrete construction), and to fasten other members and appurtenances.

4.22 General Requirements

4.22.1 The design of studs shall be suitable for arc welding to steel members with automatically timed stud welding equipment. The type, size or diameter, and length of stud shall be as specified by the drawings, specifications, or special provisions as approved by the Engineer. (See Fig. 4.22.1 for dimensions and tolerances of standard-type shear connectors.)

4.22.2 An arc shield (ferrule) of heat-resistant ceramic or other suitable material shall be furnished with each stud.

Table 4.16
All-weld-metal tension test requirements

Weld metal properties for joining	Tensile strength		Yield point		Elongation % in 2 in.
	ksi	MPa	ksi	MPa	
ASTM A36	60	415	36	250	24*
ASTM A242 or A441 Thickness (T) limitations					
T≤3/4 in. (19.0 mm)	70	485	50	345	22
3/4<T≤1-1/2 (38.1 mm)	67	460	46	315	22
1-1/2<T≤4 (102 mm)	63	435	42	290	24*
4<T≤8 (203 mm)	60	415	40	275	24*
ASTM A572					
Grade 42	60	415	42	290	24*
Grade 45	60	415	45	310	22
Grade 50	65	450	50	345	21
Grade 55	70	485	55	380	20
Grade 60	75	515	60	415	18
Grade 65	80	550	65	450	17
ASTM A588 Thickness (T) limitations					
T≤4 in. (102 mm)	70	485	50	345	21*
4 in.<T≤5 in. (127 mm)	67	460	46	315	21*
5 in. <T≤8 in. (203 mm)	63	435	42	290	21*

*A deduction in specified percentage of elongation of 0.5 percent shall be made for each 1/2 in. (12.7 mm) increase above 3-1/2 in. (88.9 mm). This deduction shall not exceed 3.0 percent.

4.22.3 A suitable deoxidizing and arc-stabilizing flux for welding shall be furnished with each stud of 5/16 in. (8.0 mm) diameter or larger. Studs less than 5/16 in. in diameter may be furnished with or without flux.

4.22.4 Only studs with qualified stud bases shall be used. A stud base, to be qualified, shall have passed the tests prescribed in 4.27. The arc shield used in production shall be the same as used in the qualification tests. Qualification of stud bases in accordance with 4.27 shall be at no expense to the owner.

4.22.5 Finish shall be produced by cold heading, cold rolling, or machining. Finished studs shall be of uniform quality and condition, free of injurious laps, fins, seams, cracks, twists, bends, or other injurious discontinuities. A stud with cracks or bursts deeper than one-half of the distance from the periphery of the head to the shank may be cause for rejection.[16]

16. Heads of shear connectors or anchor studs are subject to cracks or bursts, which are names for the same thing. Cracks or bursts designate an abrupt interruption of the periphery of the stud head by radial separation of the metal. Such interruptions do not adversely affect the structural strength, corrosion resistance, or other functional requirements of steel connectors or anchor studs.

4.22.6 Only studs qualified under 4.27 shall be used. When requested by the Engineer, the contractor shall provide the following information.

4.22.6.1 A description of the stud and arc shield.

4.22.6.2 Certification from the manufacturer that the stud base is qualified as specified in 4.22.4. Qualification test data shall be retained in the files of the manufacturer. Copies of the data shall be furnished by the contractor or fabricator on written request of the Engineer.

4.23 Mechanical Requirements

4.23.1 Studs shall be made from cold drawn bar stock conforming to the requirements of ASTM A108, Specification for Cold Finished Carbon Steel Bars and Shafting, Grades 1010 through 1020, either semi- or fully-killed.

4.23.1.1 Tensile requirements of shear connector studs, as determined by tests of bar stock after drawing or of full diameter finished studs, at the manufacturer's option, shall conform to the following:

Tensile strength, psi min	60 000 (415 MPa)
Elongation in 2 in. (50.8 mm), % min	20
Reduction of area, % min	50

4.23.1.2 Studs other than shear connectors shall have a minimum tensile strength of 55 000 psi (380 MPa) and a minimum elongation of 20% in 2 in. (50.8 mm). Tests may be made on bar stock after drawing or on full diameter finished studs, at the manufacturer's option.

4.23.2 Tensile requirements shall be determined in accordance with the applicable sections of ASTM A370, Mechanical Testing of Steel Products. When the tensile requirements of 4.23.1.1 are determined from finished studs, the tension tests may be made on studs welded to test plates of ASTM A36 steel, using a test fixture similar

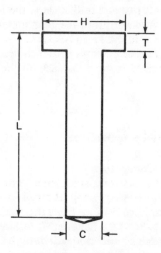

Note: L = manufactured length − length specified by engineer plus burn-off

Standard dimensions, in.			
Shank diameter (C)	Length tolerances (L)	Head diameter (H)	Minimum head height (T)
1/2 $^{+0.000}_{-0.010}$	± 1/16	1 ± 1/64	9/32
5/8 $^{+0.000}_{-0.010}$	± 1/16	1-1/4 ± 1/64	9/32
3/4 $^{+0.000}_{-0.015}$	± 1/16	1-1/4 ± 1/64	3/8
7/8 $^{+0.000}_{-0.015}$	± 1/16	1-3/8 ± 1/64	3/8
Standard dimensions, mm			
12.7 $^{+0.00}_{-0.25}$	± 1.6	25.4 ± 0.4	7.1
15.9 $^{+0.00}_{-0.25}$	± 1.6	31.7 ± 0.4	7.1
19.0 $^{+0.00}_{-0.38}$	± 1.6	31.7 ± 0.4	9.5
22.1 $^{+0.00}_{-0.38}$	± 1.6	34.9 ± 0.4	9.5

Fig. 4.22.1—Dimensions and tolerances of standard-type shear connectors

to that shown in Fig. 4.23.2. When the tensile requirements of 4.23.1.2 are determined from finished studs, the ends of the studs may be gripped in the jaws of a tension testing machine. Plates of adequate size may be fillet welded to the unwelded end for studs without heads. If fracture occurs outside the middle half of the gage length, the test shall be repeated.

4.23.3 Upon request by the Engineer, the contractor shall furnish
(1) The stud manufacturer's certification that the studs, as delivered, conform to the applicable requirements of 4.22 and 4.23.
(2) Certified copies of the stud manufacturer's test reports covering the last completed set of in-plant quality control mechanical tests, required by 4.23, for each stock size delivered. These tests shall be made using either finished studs or steel bars for studs of diameters to be furnished under the contract. The quality control tests shall have been made within the six month period before delivery of the studs.

4.23.3.1 When quality control tests are not available, the contractor shall furnish mechanical test reports conforming to the requirements of 4.23. The mechanical tests shall be on either finished studs or steel bars for studs of diameters to be delivered and selected from material provided by the manufacturer of the studs. The number of tests to be performed shall be specified by the Engineer.

4.23.4 The Engineer may, at the contractor's expense, select studs of each type and size used under the contract as necessary for checking the requirements of 4.22 and 4.23. These check tests shall be at the owner's expense.

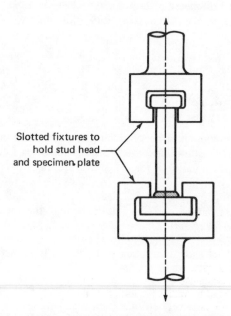

Slotted fixtures to hold stud head and specimen plate

Fig. 4.23.2—Typical tension test fixture

4.24 Workmanship

4.24.1 Studs shall be welded to steel members with automatically timed stud welding equipment connected to a suitable power source.

4.24.2 If two or more stud welding guns are to be operated from the same power source, they shall be interlocked so that only one gun can operate at a time, and so that the power source has fully recovered from making one weld before another weld is started.

4.24.3 While in operation, the welding gun shall be held in position without movement until the weld metal has solidified.

4.24.4 At the time of welding, the studs shall be free from rust, rust pits, scale, oil, or other deleterious matter that would adversely affect the welding operation.

4.24.5 The stud base shall not be painted, galvanized, or cadmium-plated prior to welding.

4.24.6 The areas on the member to which the studs are to be welded shall be free of scale, rust, or other injurious material to the extent necessary to obtain satisfactory welds. These areas may be cleaned by wire brushing, peening, prick-punching, or grinding.[17]

4.24.7 Welding shall not be done when the base metal temperature is below 0° F (−18° C) or when the surface is wet or exposed to falling rain or snow. When the temperature of the base metal is below 32° F (0° C), one stud in each 100 studs welded shall be tested by the methods specified in 4.26.1 and 4.26.2, as applicable, in addition to the first two tested, as specified in 4.25.1 and 4.25.2.

4.24.8 Longitudinal and lateral spacings of stud shear connectors with respect to each other and to edges of beam or girder flanges may vary a maximum of 1 in. (25.4 mm) from the location shown in the drawings, provided the adjacent studs are not closer than 2-1/2 in. (63.5 mm) center-to-center. The minimum distance from the edge of a stud base to the edge of a flange shall be the diameter of the stud plus 1/8 in. (3.2 mm) but preferably not less than 1-1/2 in. (38.1 mm). Other types of studs shall be so located as to permit a workmanlike assembly of attachments without alterations or reaming.

4.24.9 After welding, arc shields shall be broken free from shear connectors and anchor studs and, where practicable, from all other studs.

4.24.10 The studs, after welding, shall be free of any discontinuities or substances that would interfere with their intended function. However, nonfusion on the vertical leg of the flash[18] and small shrink fissures are acceptable.[19]

4.24.11 At the option of the contractor, studs may be fillet welded by the shielded metal arc process, provided the following requirements are met.

4.24.11.1 The fillet weld size shall be a minimum of 5/16 in. (8.0 mm).

4.24.11.2 Welding shall be done with low-hydrogen electrodes 5/32 or 3/16 in. (4.0 or 4.8 mm) in diameter.

4.24.11.3 The stud base shall be prepared so that the outside circumference of the stud fits tightly against the base metal.

4.24.11.4 All rust and mill scale at the location of the stud shall be removed from the base metal by grinding. The end of the stud shall also be clean.

4.24.11.5 The base metal to which studs are welded shall be preheated in accordance with the requirements of Table 4.2.

4.25 Quality Control

4.25.1 Shear Connectors

4.25.1.1 The first two stud shear connectors welded on each member, after being allowed to cool, shall be bent to an angle of 30 deg from their original axes by striking the studs with a hammer. If failure occurs in the weld zone of either stud, the procedure shall be corrected and two more studs shall be welded to the member and tested. If either of the second two studs fails, additional welding shall be continued on separate plates until two consecutive studs are tested and found to be satisfactory. Two consecutive studs shall then be welded to the member, tested, and found to be satisfactory before any more production studs are welded to the member.

4.25.1.2 For members having less than 20 stud shear connectors, the stud welding procedure may be tested at the start of each day's production welding period[20] in lieu

---

17. Extreme care should be exercised when welding through metal decking.

18. The fillet weld profiles shown in Fig. 3.6 do not apply to the flash of automatically timed stud welds.

19. The expelled metal around the base of the stud is designated as flash in accordance with the definition of flash in AWS A3.0, Terms and Definitions, and Appendix I of this code. It is not a fillet weld such as those formed by conventional arc welding. The expelled metal, which is excess to the weld required for strength, is not detrimental but, on the contrary, is essential to provide a good weld. The containment of this excess molten metal around a welded stud by the ferrule (arc shield) assists in securing sound fusion of the entire cross section of the stud base. The stud weld flash may have nonfusion in its vertical leg and overlap on its horizontal leg; and it may contain occasional small-shrink fissures or discontinuities that usually form at the top of the weld flash with essentially radial or longitudinal orientation, or both, to the axis of the stud. Such nonfusion on the vertical leg of the flash and small-shrink fissures are acceptable.

20. A new production period begins with the welding of a given size and type stud with a given welding procedure or with the beginning of each day's production.

of testing in accordance with 4.25.1.1. Before use in production, each welding unit shall be used to weld two stud shear connectors to separate test material in the same general position (flat, vertical, overhead, sloping) and of similar thickness. After being allowed to cool, they shall be bent as described in 4.25.1.1. If failure occurs, the procedure shall be corrected and two consecutive studs shall be welded to the test material, tested, and found to be satisfactory before any production studs are welded to the member.

4.25.1.3 The foregoing testing shall be performed after any change in the welding procedure.

4.25.1.4 If failure occurs in the stud shank, an investigation shall be made to ascertain and correct the cause before more studs are welded.

4.25.2 Applications Other Than Shear Connectors. Before starting the welding operations or at the request of the Engineer, two stud connectors shall be welded to separate material in the same general position (flat, vertical, overhead, sloping) and of thickness and material similar to the member. After being allowed to cool, each stud shall be bent to an angle of 30 deg from its original axis by striking the stud with a hammer. If failure occurs in the weld zone of either stud, the procedure shall be corrected and two successive studs successfully welded and tested before any studs are welded to the member. The foregoing testing shall be performed after any change in the welding procedure. If failure occurs in the stud shank, an investigation shall be made to ascertain and correct the cause before more studs are welded.

4.25.3 Studs on which a full 360 deg flash[18] is not obtained may, at the option of the stud welding contractor, be repaired by adding 5/16 in. (8.0 mm) minimum fillet weld in place of the missing flash.[18] The shielded metal arc process with low-hydrogen electrodes, 5/32 or 3/16 in. (4.0 or 4.8 mm) in diameter, shall be used in accordance with the requirements of this code. The repair weld shall extend at least 3/8 in. (9.5 mm) beyond each end of the discontinuity being repaired.

4.25.3.1 For studs having a shank diameter 7/16 in. (11 mm) or less, the use of smaller diameter electrodes is permissible provided they are low-hydrogen type.

4.25.4 Operator Qualification. The initial test required by 4.25.1 or 4.25.2 shall also serve to qualify the stud welding operator.

4.25.4.1 Before any production studs are welded by an operator not involved in the initial procedure qualification, the operator shall have the first two welded studs tested in accordance with the provisions of 4.25.1 or 4.25.2, as applicable. When the two welded studs have been tested and found satisfactory, the operator may then weld production studs.

4.25.5 If an unacceptable stud has been removed from a component subjected to tensile stresses, then the area from which the stud was removed shall be made smooth and flush. Where in such areas base metal has been pulled out in the course of stud removal, shielded metal arc welding with low-hydrogen electrodes in accordance with the requirements of this code shall be used to fill the pockets and the weld surface shall be ground flush. In compression areas of members, if stud failures are confined to shanks or fusion zones of studs, a new stud may be welded adjacent to each unacceptable area in lieu of repair and replacement on the existing weld area (see 4.24.8). If metal is torn from the base metal of such areas, the repair provisions shall be the same as for tension areas except that when the depth of discontinuity is less than 1/8 in. (3.2 mm) and 7% of the base metal thickness the discontinuity may be faired by grinding in lieu of filling the unacceptable area with weld metal. Where a replacement stud is to be placed in the unacceptable area, the just-mentioned repair shall be made prior to welding the replacement stud. Replacement shear connector studs shall be tested by bending to an angle of 15 deg from their original axes. The areas of components exposed to view in completed structures shall be made smooth and flush where a stud has been removed.

4.26 Inspection Requirements

4.26.1 If a visual inspection reveals any stud shear connector that does not show a full 360 deg flash,[18] any stud that has been repaired by welding, or any stud in which the reduction in length due to welding is less than normal shall be struck with a hammer and bent to an angle of 15 deg from its original axis. For studs showing less than a 360 deg weld fillet, the direction of bending shall be opposite to the missing weld fillet. Studs that crack in the weld, the base metal, or the shank under inspection or subsequent straightening shall be replaced (see 4.26.4).

Nonfusion (on the vertical leg of the flash)[18] and small-shrink fissures are acceptable.[19]

4.26.2 For studs other than shear connectors, at least one stud in every 100 shall be bent to an angle of 15 deg from its original axis by striking with a hammer. If threaded, the stud shall be torque-tested with a calibrated torque wrench to the value given in Fig. 4.26.2 for the diameter and thread of the stud, in a device similar to that shown in Fig. 4.26.2. If the stud fails, the procedures shall be checked in accordance with 4.25.2, and two more of the existing studs shall be bent or torque-tested. If either of these two studs fails, all of the studs represented by the tests shall be torque-tested, bend-tested, or rejected. For critical structural connections, the Engineer shall designate the type and extent of additional inspection in the contract.

Nonfusion (on the vertical leg of the flash)[18] and small-shrink fissures are acceptable.[19]

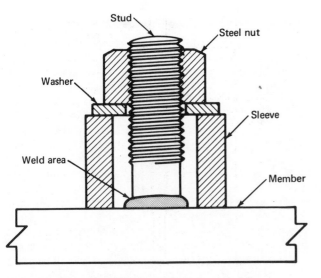

Note: The dimensions are appropriate to the size of the stud. The threads of the stud shall be clean and free of lubricant other than the residue of cutting oil.

Required torque for testing threaded studs				
Nominal diameter of studs		Threads per inch & series designated	Testing torque	
in.	mm		ft-lb	J
1/4	6.4	28 UNF	5.0	6.8
1/4		20 UNC	4.2	5.7
5/16	7.9	24 UNF	9.5	12.9
5/16		18 UNC	8.6	11.7
3/8	9.5	24 UNF	17.0	23.0
3/8		16 UNC	15.0	20.3
7/16	11.1	20 UNF	27.0	36.6
7/16		14 UNC	24.0	32.5
1/2	12.7	20 UNF	42.0	57.0
1/2		13 UNC	37.0	50.2
9/16	14.3	18 UNF	60.0	81.4
9/16		12 UNC	54.0	73.2
5/8	15.9	18 UNF	84.0	114.0
5/8		11 UNC	74.0	100.0
3/4	19.0	16 UNF	147.0	200.0
3/4		10 UNC	132.0	180.0
7/8	22.2	14 UNF	234.0	320.0
7/8		9 UNC	212.0	285.0
1.0	25.4	12 UNF	348.0	470.0
1.0		8 UNC	318.0	430.0

Fig. 4.26.2—Torque testing arrangement and table of testing torques

4.26.3 The Engineer's Inspector, where conditions warrant, may select a reasonable number of additional studs to be subjected to the tests specified in 4.26.1 and 4.26.2.

4.26.4 The bent stud shear connectors and concrete anchors that show no sign of failure shall be acceptable for use and left in the bent position if no portion of the stud is less than 1 in. (25.4 mm) from a proposed concrete surface. All required bending and straightening shall be done, without heating, before completion of the stud welding operation on the job, except as otherwise provided in the contract.

4.26.5 If, in the judgment of the Engineer, studs welded during the progress of the work are not in accordance with code provisions, as indicated by inspection and testing, corrective action shall be required of the contractor. At his own expense, the contractor shall make the changes (such as welding procedure, welding equipment, and stud base) necessary to ensure that studs subsequently welded will meet code requirements.

4.26.6 At the option and the expense of the owner, the contractor may be required at any time to submit studs of the types used under the contract for check qualification in accordance with the procedures of 4.27.

4.27 Stud Base Qualification Requirements

4.27.1 Purpose. The purpose of these requirements is to prescribe tests for the stud manufacturer's certification of a stud base for welding under shop or field conditions.

4.27.2 Responsibility for Tests. The stud manufacturer shall be responsible for the performance of the qualification tests. These tests may be performed by a testing agency satisfactory to the Engineer. The agency performing the tests shall submit a certified report to the manufacturer of the studs giving procedures and results for all tests including the information listed under 4.27.10.

4.27.3 Extent of Qualification. Qualification of a stud base shall constitute qualification of stud bases with the same geometry, flux, and arc shield, having the same diameter and diameters that are smaller by less than 1/8 in. (3 mm).[21] A stud base qualified with an approved grade of ASTM A108 steel shall constitute qualification for all other approved grades of A108 steel (see 4.23.1) provided that all other provisions stated herein are complied with.

4.27.4 Duration of Qualification. A size of stud base with arc shield, once qualified, is considered qualified until the stud manufacturer makes any change in the stud base geometry, material, flux, or arc shield which affects the welding characteristics.

---

21. For example, qualification of a 3/4 in. (19 mm) shank diameter stud base does not constitute qualification for a 5/8 in. (16 mm) shank diameter stud, but would constitute qualification for a stud base having a shank diameter of 41/64 in. (16 mm).

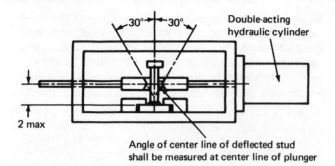

Double-acting hydraulic cylinder

30° — 30°

2 max

Angle of center line of deflected stud shall be measured at center line of plunger

Notes:

1. Fixture holds specimen and stud is bent 30° alternately in opposite directions.

2. Load can be applied with hydraulic cylinder (shown) or fixture adapted for use with tension test machine.

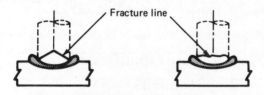

Fracture line

Typical fractures in shank of stud

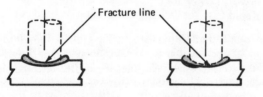

Fracture line

Note: Fracture in weld near stud fillet remains on plate

Note: Fracture through flash torn from plate

Typical weld failures

Fig. 4.27.7.2A—Bend testing device

4.27.5 Preparation of Specimens

4.27.5.1 Test specimens shall be prepared by welding representative studs to suitable specimen plates of ASTM A36 steel. Tests for threaded studs shall be on blanks (studs without threads).

4.27.5.2 Studs shall be welded with power source, welding gun, and automatically controlled equipment as recommended by the stud manufacturer. Welding voltage, current, and time (see 4.27.6) shall be measured and recorded for each specimen. Lift and plunge shall be at the optimum setting as recommended by the manufacturer.

4.27.6 Number of Test Specimens

4.27.6.1 Thirty test specimens shall be welded consecutively with constant optimum time but with current 10% above optimum. Optimum current and time shall be the midpoint of the range normally recommended by the manufacturer for production welding.

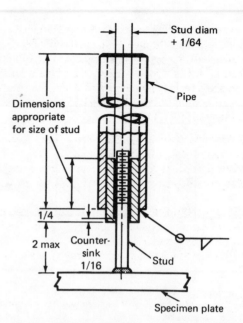

Stud diam + 1/64

Pipe

Dimensions appropriate for size of stud

1/4

2 max

Countersink 1/16

Stud

Specimen plate

Fig. 4.27.7.2B—Suggested type of device for qualification testing of small studs

4.27.6.2 Thirty test specimens shall be welded consecutively with constant optimum time but with current 10% below optimum.

4.27.7 Tests

4.27.7.1 Tension Tests. Ten of the specimens welded in accordance with 4.27.6.1 and ten in accordance with 4.27.6.2 shall be subjected to a tension test in a fixture similar to that shown in Fig. 4.23.2, except that studs without heads may be gripped on the unwelded end in the jaws of the tension testing machine. A stud base shall be considered as qualified if all test specimens have a tensile strength equal to or above the minimum specified in 4.23.1.1.

4.27.7.2 Bend Tests. Twenty of the specimens welded in accordance with 4.27.6.1 and twenty in accordance with 4.27.6.2 shall be bend-tested by being bent alternately 30 deg from their original axes in opposite directions until failure occurs. Studs shall be bent in a bend-testing device as shown in Fig. 4.27.7.2A, except that studs less than 1/2 in. (12.7 mm) diam, optionally, may be bent using a device as shown in Fig. 4.27.7.2B. A stud base shall be considered as qualified if, on all test specimens, fracture occurs in the plate material or shank of the stud and not in the weld or heat-affected zone.

4.27.8 Retests. If failure occurs in a weld or heat-affected zone in any of the bend test groups of 4.27.7.2 or at less than the specified minimum tensile strength of the stud in any of the tension test groups of 4.27.7.1, a new test group (specified in 4.27.6.1 or 4.27.6.2 as applicable) shall be prepared and tested. If such failures repeat, the stud base shall fail to qualify.

4.27.9 Acceptance. For a manufacturer's stud base and arc shield combination to be qualified, each stud of each group of 30 studs shall, by test or retest, meet the requirements prescribed in 4.27.7. Qualification of a given diameter of stud base shall be considered qualification for stud bases of the same nominal diameter (see 4.27.3), stud base geometry, material, flux, and arc shield.

4.27.10 Manufacturer's Qualification Test Data. The data shall include the following.

4.27.10.1 Drawings that show shapes and dimensions with tolerances of studs, arc shields, and, if used, sheet flux.

4.27.10.2 A complete description of materials used in the studs, including the quantity and type of flux, and a description of the arc shields.

4.27.10.3 Certified results of laboratory tests required by 4.27.

Part G
Plug and Slot Welds

4.28 Plug Welds

The technique used to make plug welds when using shielded metal arc welding, gas metal arc welding (except short-circuiting transfer), and flux cored arc welding processes shall be as follows.

4.28.1 For welds to be made in the flat position, each shall be deposited around the root of the joint and then deposited along a spiral path to the center of the hole, fusing and depositing a layer of weld metal in the root and bottom of the joint. The arc is then carried to the periphery of the hole and the procedure repeated, fusing and depositing successive layers to fill the hole to the required depth. The slag covering the weld metal should be kept molten until the weld is finished. If the arc is broken or the slag is allowed to cool, the slag must be completely removed before restarting the weld.

4.28.2 For welds to be made in the vertical position, the arc is started at the root of the joint, at the lower side of the hole, and is carried upward, fusing into the face of the inner plate and to the side of the hole. The arc is stopped at the top of the hole, the slag is cleaned off, and the process is repeated on the opposite side of the hole. After cleaning slag from the weld, other layers should be similarly deposited to fill the hole to the required depth.

4.28.3 For welds to be made in the overhead position, the procedure is the same as for the flat position, except that the slag should be allowed to cool and should be completely removed after depositing each successive bead until the hole is filled to the required depth.

4.29 Slot Welds

Slot welds shall be made using techniques similar to those specified in 4.28 for plug welds, except that if the length of the slot exceeds three times the width, or if the slot extends to the edge of the part, the technique requirements of 4.28.3 shall apply.

5. Qualification

Part A
General Requirements

5.1 Approved Procedures

5.1.1 Joint welding procedures that conform in all respects to the pertinent provisions of Section 2, Design of Welded Connections; Section 3, Workmanship; Section 4, Technique; Section 8, Design of New Buildings; Section 9, Design of New Bridges; or Section 10, Design of New Tubular Structures shall be deemed prequalified and are exempt from tests or qualifications. Prequalified joint welding procedures shall meet all of the requirements listed in Appendix E, Table E1—Mandatory Code Requirements for Prequalified Joint Welding Procedures.[22]

Note: The use of a prequalified joint welding procedure is not intended as a substitute for engineering judgment in the suitability of application of these joint welding procedures to a welded assembly or connection.

5.1.2 All prequalified joint welding procedures to be used shall be prepared by the manufacturer, fabricator, or contractor as written procedure specifications and shall be available to those authorized to examine them. A suggested form showing the information required in the procedure specification is given in Appendix E.

5.1.3 A combination of qualified or prequalified joint welding procedures may be used without qualification provided the limitation of essential variables applicable to each process is observed.

5.2 Other Procedures

Except for the procedures exempted in 5.1, joint welding procedures which are to be employed in executing contract work under this code shall be qualified prior to use, to the satisfaction of the Engineer, by tests as prescribed in Part B of this section. The Engineer, at his discretion, may accept evidence of previous qualification of the joint welding procedures to be employed.[23, 24]

5.3 Welders, Welding Operators, and Tackers

5.3.1 All welders, welding operators, and tackers to be employed under this code shall have been qualified by tests as prescribed in Parts C, D, and E of this section.

5.3.2 Radiographic examination of a welder's or welding operator's qualification test plate or test pipe may be made in lieu of the guided-bend tests prescribed in Parts C and D of this section.

5.4 Qualification Responsibility

5.4.1 Each manufacturer or contractor shall conduct the tests required by this code to qualify the welding procedures.

5.4.2 The Engineer, at his discretion, may accept evidence of previous qualification of welders, welding operators, and tackers to be employed.[23]

---

22. All of the provisions listed in Appendix E, Table E1, must be observed for a joint welding procedure to be deemed prequalified. All other provisions of the code, even though they do not relate to joint welding procedures, are also mandatory.

23. The Engineer should accept properly documented evidence of previous qualification tests.

24. Only the requirements listed in Appendix E, Table E2, Code Requirements That May be Changed by Procedure Qualification Tests, may be varied when the procedure is qualified by tests. No other code requirement may be changed by procedure qualification.

Part B
Procedure Qualification[25]

5.5 Limitation of Variables

5.5.1 When necessary to establish a welding procedure by qualification as required by 5.2 or by contract specifications, the following rules apply and the procedure shall be recorded by the manufacturer or contractor as a procedure specification.

5.5.1.1 Qualification of a welding procedure established with a base metal included in 10.2, not listed in 5.5.1.2, and having a minimum specified yield point less than 50 000 psi (345 MPa) shall qualify the procedure for welding any other base metal or combination of those base metals included in 10.2 that has a minimum specified yield point equal to or less than that of the base metal used in the test.

5.5.1.2 Qualification of a welding procedure established with one of the materials listed in Group II of Table 4.1.1 shall be considered as procedure qualification for welding any other steel in this group, combinations within this group, or any steel included in 10.2 that has a lower minimum specified yield point.

5.5.1.3 Qualification of a welding procedure established with a base metal included in 10.2 having a minimum specified yield strength greater than 50 000 psi (345 MPa) shall qualify the procedure for welding only base metals of the same material specification and grade or type, having the same minimum specified yield strength as the base metal tested, reduction in yield strength for increase in material thickness excepted. For example, a procedure qualified with a 1 in. (25.4 mm) thick 100 000 psi (690 MPa) yield strength base metal also qualifies for a 3 in. (76.2 mm) thick 90 000 psi (620 MPa) yield strength base metal of the same material specification.

5.5.1.4 Qualification of a welding procedure established with a combination of base metals (included in 10.2) having different minimum specified yield strengths, one of which is greater than 50 000 psi (345 MPa), shall qualify the procedure for welding that high yield-strength base metal to any other of those base metals having a minimum specified yield strength equal to or less than that of the lower strength base metal in the test.

5.5.1.5 In preparing the procedure specification, the manufacturer or contractor shall report the specific values for the essential variables that are specified in 5.5. The suggested form for showing the information required in the procedure specification is given in Appendix E.

5.5.2 The changes set forth in 5.5.2.1 through 5.5.2.5

shall be considered essential changes in a welding procedure and shall require establishing a new procedure by qualification. When a combination of welding processes is used, the variables applicable to each process shall apply.

5.5.2.1 Shielded Metal Arc Welding

(1) A change increasing filler metal strength level (a change from E70XX to E80XX, for example, but not vice versa).

(2) A change from a low-hydrogen type electrode to a non-low-hydrogen type of electrode, but not vice versa.

(3) An increase of electrode diameter by more than 1/32 in. (1 mm) over that used in the procedure qualification.

(4) A change of electrode amperage and voltage values that is not within the ranges recommended by the electrode manufacturer.[26]

(5) For a specified groove, a change of more than ±25% in the specified number of passes. If the area of the groove is changed, it is permissible to change the number of passes in proportion to the area.

(6) A change in position in which welding is done as defined in 5.8.

(7) A change in the type of groove (a change from a V- to a U-groove, for example).

(8) A change exceeding tolerances of 2.9, 2.10, or 10.13 in the shape of any one type of groove involving

    (a) A decrease in the included angle of the groove

    (b) A decrease in the root opening of the groove

    (c) An increase in the root face of the groove

    (d) The omission, but not inclusion, of backing material

(9) A decrease of more than 25° F (13.9° C) in the minimum specified preheat or interpass temperature.[27]

(10) In vertical welding, a change in the progression specified for any pass from upward to downward or vice versa.

(11) The omission, but not the inclusion, of back gouging.

5.5.2.2 Submerged Arc Welding

(1) A change in electrode and flux combination not covered by AWS A5.17 or A5.23.

(2) A change increasing filler metal strength level (from Grade F80 to Grade F90, for example, but not vice versa).

(3) A change in electrode diameter when using an alloy flux.[28]

---

25. Welding procedures for processes listed in 1.3 and qualified in accordance with the requirements of previous editions of this code shall be considered to have qualified under the tests prescribed herein subject to the limitation of variables in 5.5. Any requalifications or new qualifications shall be made in accordance with the requirements of this edition.

26. When welding quenched and tempered steel, any change within the limitation of variables shall not increase the heat input beyond the steel producer's recommendations.

27. The temperature may fall more than 25° F (13.9° C) below the minimum specified, provided (1) the provisions of 3.4.7 and Table 4.2 are complied with, and (2) the work shall be at the specified minimum temperature at the time of subsequent welding.

28. An alloy flux is defined as a flux upon which the alloy content of the weld metal is largely dependent.

(4) A change in the number of electrodes used.

(5) A change in the type of current (ac or dc) or polarity when welding quenched and tempered steel or when using an alloy flux.

(6) A change of more than 10% above or below the specified mean amperage for each electrode diameter used.[26]

(7) A change of more than 7% above or below the specified mean arc voltage for each diameter electrode used.[26]

(8) A change of more than 15% above or below the specified mean travel speed.[26]

(9) A change of more than 10%, or 1/8 in. (3.2 mm), whichever is greater, in the longitudinal spacing of the arcs.

(10) A change of more than 10%, or 1/16 in. (1.6 mm) whichever is greater, in the lateral spacing of the arcs.

(11) A change of more than ± 10 deg in the angular position of any parallel electrode.

(12) A change in the angle of electrodes in machine or automatic welding of more than

(a) ± 3 deg in the direction of travel

(b) ± 5 deg normal to the direction of travel

(13) For a specified groove, a change of more than ± 25% in the specified number of passes. If the area of the groove is changed, it is permissible to change the number of passes in proportion to the area.

(14) A change in position in which welding is done as defined in 5.8.

(15) A change in the type of groove (a change from a V- to a U-groove, for example).

(16) A change, exceeding tolerances of 2.9, 2.10, and 3.3.4, in the shape of any one type of groove involving

(a) A decrease in the included angle of the groove

(b) A decrease in the root opening of the groove

(c) An increase in the root face of the groove

(d) The omission, but not inclusion, of backing material

(17) A decrease of more than 25° F (13.9° C) in the minimum specified preheat or interpass temperature.[27]

(18) An increase in the diameter of the electrode used over that called for in the procedure specification.

(19) The addition or deletion of supplemental powdered or granular filler metal or cut wire.

(20) An increase in the amount of supplemental powdered or granular filler metal or cut wire.

(21) If the alloy content of the weld metal is largely dependent on the composition of supplemental powdered filler metal, any change in any part of the joint welding procedure which would result in important alloying elements in the weld metal not meeting the chemical requirements given in the welding procedure specification.

(22) The omission, but not the inclusion, of back gouging.

5.5.2.3 Gas Metal Arc Welding

(1) A change in electrode and method of shielding not covered by AWS A5.18.

(2) A change increasing filler metal strength level (from Grade E70S to Grade E80S, for example, but not vice versa).

(3) A change in electrode diameter.

(4) A change in the number of electrodes used.

(5) A change from a single gas to any other single gas or to a mixture of gases or a change in specified percentage composition of gas mixture not covered by AWS A5.18.

(6) A change of more than 10% above or below the specified mean amperage for each diameter electrode used.[26]

(7) A change of more than 7% above or below the specified mean arc voltage for each diameter electrode used.[26]

(8) A change of more than 10% above or below the specified mean travel speed.[26]

(9) An increase of 25% or more or a decrease of 10% or more in the rate of flow of shielding gas or mixture.

(10) For a specified groove, a change of more than ± 25% in the specified number of passes. If the area of the groove is changed, it is permissible to change the number of passes in proportion to the area.

(11) A change in the position in which welding is done, as defined in 5.8.

(12) A change in the type of groove (a change from a V- to U-groove, for example).

(13) A change, exceeding tolerances in 2.9, 2.10, or 10.13; and 3.3.4; or 10.14.3, in the shape of any one type of groove involving

(a) A decrease in the included angle of the groove

(b) A decrease in the root opening of the groove

(c) An increase in the root face of the groove

(d) The omission, but not inclusion, of backing material

(14) A decrease of more than 25° F (13.9° C) in the minimum specified preheat or interpass temperature.[27]

(15) In vertical welding, a change in the progression specified for any pass from upward to downward, or vice versa.

(16) A change in type of welding current (ac or dc), polarity, or mode of metal transfer across the arc.

(17) The omission, but not the inclusion, of back gouging.

5.5.2.4 Flux Cored Arc Welding

(1) A change in electrode and method of shielding not covered by AWS A5.20.

(2) A change increasing filler metal strength level (from Grade E70T to E80T, for example, but not vice versa).

(3) An increase in the diameter of electrode used over that called for in the procedure specification.

(4) A change in the number of electrodes used.

(5) A change from a single gas to any other single gas or to a mixture of gases or a change in specified percentage composition of gas mixture not covered by AWS A5.20.

(6) A change of more than 10% above or below the specified mean amperage for each size electrode used.[26]

(7) A change of more than 7% above or below the specified mean arc voltage for each size electrode used.[26]

(8) A change of more than 10% above or below the specified mean travel speed.[26]

(9) An increase of 25% or more or a decrease of 10% or more in the rate of flow of shielding gas or mixture.

(10) For a specified groove, a change of more than ± 25% in the specified number of passes. If the area of the groove is changed, it is permissible to change the number of passes in proportion to the area.

(11) A change in the position in which welding is done as defined in 5.8.

(12) A change in the type of groove (a change from a V- to a U-groove, for example).

(13) A change, exceeding tolerances in 2.9, 2.10, or 10.13; and 3.3.4 or 10.14.3, in the shape of any one type of groove involving

(a) A decrease in the included angle of the groove

(b) A decrease in the root opening of the groove

(c) An increase in the root face of the groove

(d) The omission, but not inclusion, of backing material

(14) A decrease of more than 25° F (13.9° C) in the minimum specified preheat or interpass temperature.[27]

(15) In vertical welding, a change in the progression specified for any pass from upward to downward or vice versa.

(16) A change in type of welding current (ac or dc), polarity, or mode of metal transfer across the arc.

(17) The omission, but not the inclusion, of back gouging.

5.5.2.5 Electroslag and Electrogas Welding

(1) A significant change in filler metal or consumable guide metal composition.

(2) A change in consumable guide metal core cross-sectional area exceeding 30%.

(3) A change in flux system (cored, magnetic electrode, external flux, etc.).

(4) A change in flux composition including consumable guide coating.

(5) A change in shielding gas composition of any one constituent of more than 5% of the total flow.

(6) A change in welding current exceeding 20%.

(7) A change in groove design, other than square groove, increasing groove cross-sectional area.

(8) A change in joint thickness (T) outside the limits of 0.5 T to 1.1 T, where T is the thickness used for the procedure qualification.

(9) A change in number of electrodes.

(10) A change from single pass to multiple pass or vice versa.

(11) A change to a combination with any other welding process or method.

(12) A change in postweld heat treatment.

(13) A change in design of molding shoes, either fixed or movable, from nonfusing solid to water-cooled or vice versa.

5.5.3 The following changes in a qualified electroslag or electrogas procedure shall require requalification of the procedure by radiographic or ultrasonic testing only, in accordance with the requirements of Part B or C of Section 6.

5.5.3.1 A change exceeding 1/32 in. (0.8 mm) in filler metal diameter.

5.5.3.2 A change exceeding 10 ipm (4.2 mm/s) in filler metal oscillation traverse speed.

5.5.3.3 A change in filler metal oscillation traverse dwell time exceeding 2 seconds except as necessary to compensate for variation in joint opening.

5.5.3.4 A change in filler metal oscillation traverse length which affects, by more than 1/8 in. (3.2 mm), the proximity of filler metal to the molding shoes.

5.5.3.5 A change in flux burden exceeding 30%.

5.5.3.6 A change in shielding gas flow rate exceeding 25%.

5.5.3.7 A change in design of molding shoes, either fixed or movable, as follows:

(1) Metallic to nonmetallic or vice versa

(2) Nonfusing to fusing or vice versa

(3) A reduction in any cross-sectional dimension or area of solid nonfusing shoe exceeding 25%

5.5.3.8 A change in welding position from vertical by more than 10 deg.

5.5.3.9 A change from ac to dc or vice versa, or a change in polarity for direct current.

5.5.3.10 A change in welding power volt-ampere characteristics from constant voltage to constant current or vice versa.

5.5.3.11 A change in voltage exceeding 10%.

5.5.3.12 A change exceeding 1/4 in. (6.4 mm) in square groove root opening.

5.5.3.13 A change in groove design other than square groove, reducing groove cross-sectional area.

5.5.3.14 A change in speed of vertical travel, if not an automatic function of arc length or deposition rate, exceeding 20% except as necessary to compensate for variation in joint opening.

5.6 Types of Tests and Purposes

The types of tests outlined below are to determine the mechanical properties and soundness of welded joints made under a given procedure specification. The tests used are as follows:

5.6.1 For Groove Welds

(1) Reduced-section tension test (for tensile strength)

(2) Root-bend test (for soundness)

(3) Face-bend test (for soundness)

(4) Side-bend test (for soundness)

(5) All-weld-metal test (for mechanical properties—electroslag and electrogas)

(6) Impact test (for toughness—when specified for electroslag or electrogas)

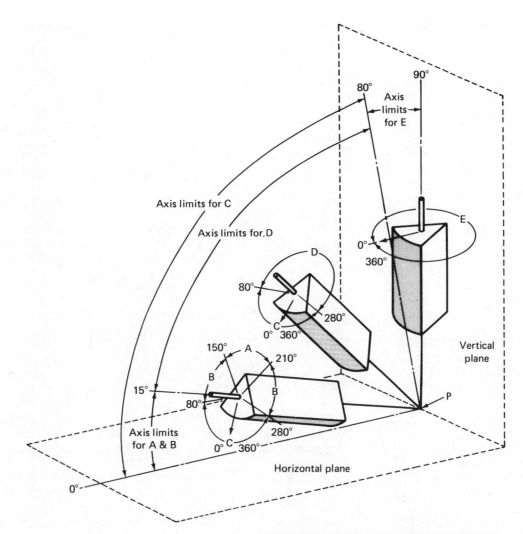

Tabulation of positions of groove welds			
Position	Diagram reference	Inclination of axis	Rotation of face
Flat	A	0° to 15°	150° to 210°
Horizontal	B	0° to 15°	80° to 150° 210° to 280°
Overhead	C	0° to 80°	0° to 80° 280° to 360°
Vertical	D E	15° to 80° 80° to 90°	80° to 280° 0° to 360°

Notes:

1. The horizontal reference plane is always taken to lie below the weld under consideration.
2. The inclination of axis is measured from the horizontal reference plane toward the vertical reference plane.
3. The angle of rotation of the face is determined by a line perpendicular to the theoretical face of the weld which passes through the axis of the weld. The reference position (0°) of rotation of the face invariably points in the direction opposite to that in which the axis angle increases. When looking at point P, the angle of rotation of the face of the weld is measured in a clockwise direction from the reference position (0°).

Fig. 5.8.1A—Positions of groove welds

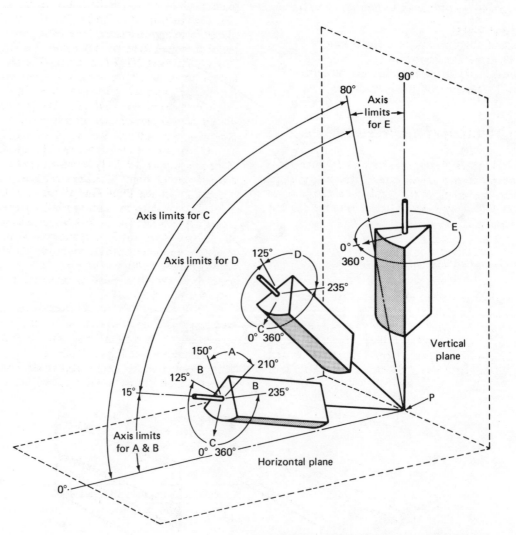

Tabulation of positions of fillet welds			
Position	Diagram reference	Inclination of axis	Rotation of face
Flat	A	0° to 15°	150° to 210°
Horizontal	B	0° to 15°	125° to 150° 210° to 235°
Overhead	C	0° to 80°	0° to 125° 235° to 360°
Vertical	D E	15° to 80° 80° to 90°	125° to 235° 0° to 360°

Fig. 5.8.1B—Positions of fillet welds

(7) Macroetch test for soundness and effective throat in partial joint penetration groove welds

(8) Radiographic or ultrasonic testing (for soundness)

5.6.2 For Fillet Welds

(1) Macroetch test for soundness and fusion

(2) Side-bend test (for soundness)

(3) All-weld-metal test (for mechanical properties)

Note: (2) and (3) for consumable verification

5.7 Base Metal and Its Preparation

The base metal and its preparation for welding shall comply with the procedure specification. For all types of welded joints, the length of the weld and dimensions of the base metal shall provide sufficient material for test specimens required by this code.

5.8 Position of Test Welds

5.8.1 All welds that will be encountered in actual construction shall be classified as: (1) flat, (2) horizontal, (3) vertical, or (4) overhead in accordance with the definitions of welding positions given in Figs. 5.8.1A and 5.8.1B. Each procedure shall be tested in the manner stated below for each position for which it is to be qualified.

5.8.1.1 Groove Plate Test Welds (illustrated in Fig. 5.8.1.1). In making the tests to qualify groove welds, test plates shall be welded in the following positions:

(1) Position 1G (Flat)—The test plates shall be placed in an approximately horizontal plane and the weld metal deposited from the upper side. See Fig. 5.8.1.1(A).

(2) Position 2G (Horizontal)—The test plates shall be placed in an approximately vertical plane with the groove approximately horizontal. See Fig. 5.8.1.1(B).

(3) Position 3G (Vertical)—The test plates shall be placed in an approximately vertical plane with the groove approximately vertical. See Fig. 5.8.1.1(C).

(4) Position 4G (Overhead)—The test plates shall be placed in an approximately horizontal plane and the weld metal deposited from the under side. See Fig. 5.8.1.1(D).

5.8.1.2 Groove Pipe Test Welds (illustrated in Fig. 5.8.1.2). In making the tests to qualify groove welds, test pipe shall be welded in the following positions:

(1) Position 1G (Pipe Horizontal Rolled)—The test pipe shall be placed with its axis horizontal and the groove approximately vertical. The pipe shall be rotated during welding so the weld metal is deposited from the upper side. See Fig. 5.8.1.2(A).

(2) Position 2G (Pipe Vertical)—The test pipe shall be placed with its axis vertical and the welding groove approximately horizontal. The pipe shall not be rotated during welding. See Fig. 5.8.1.2(B).

(3) Position 5G (Pipe Horizontal Fixed)—The test pipe shall be placed with its axis horizontal and the

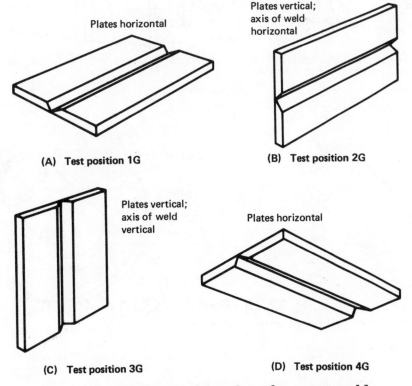

Plates horizontal

(A) Test position 1G

Plates vertical; axis of weld horizontal

(B) Test position 2G

Plates vertical; axis of weld vertical

(C) Test position 3G

Plates horizontal

(D) Test position 4G

Fig. 5.8.1.1—Positions of test plates for groove welds

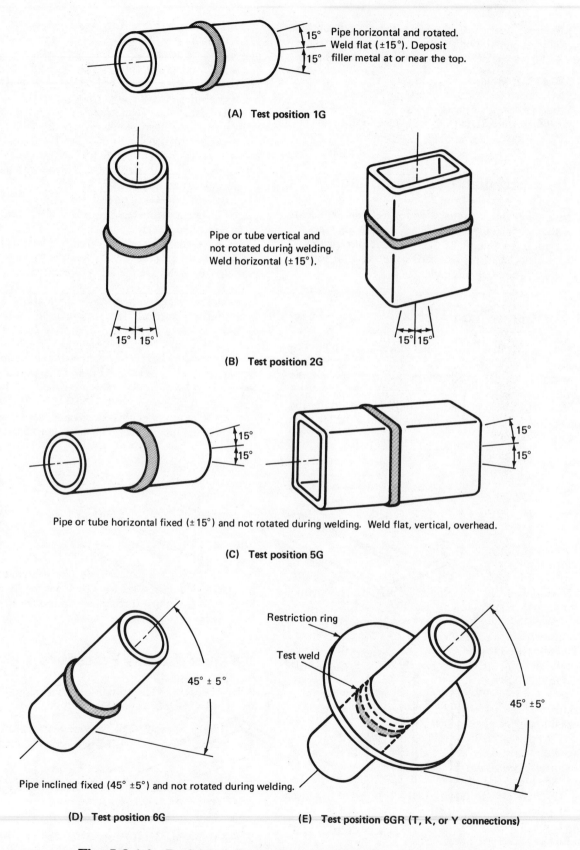

(A) Test position 1G

Pipe horizontal and rotated. Weld flat (±15°). Deposit filler metal at or near the top.

Pipe or tube vertical and not rotated during welding. Weld horizontal (±15°).

(B) Test position 2G

Pipe or tube horizontal fixed (±15°) and not rotated during welding. Weld flat, vertical, overhead.

(C) Test position 5G

Pipe inclined fixed (45° ±5°) and not rotated during welding.

(D) Test position 6G

Restriction ring

Test weld

(E) Test position 6GR (T, K, or Y connections)

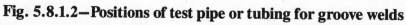

Fig. 5.8.1.2—Positions of test pipe or tubing for groove welds

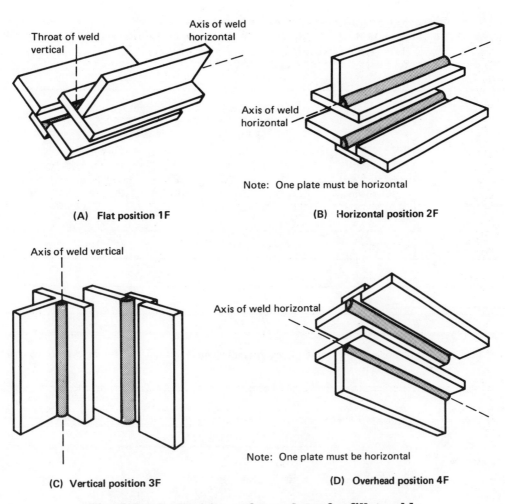

(A) Flat position 1F

(B) Horizontal position 2F

Note: One plate must be horizontal

(C) Vertical position 3F

(D) Overhead position 4F

Note: One plate must be horizontal

Fig. 5.8.1.3—Positions of test plates for fillet welds

groove approximately vertical. The pipe is not rotated during welding. See Fig. 5.8.1.2(C).

(4) Position 6G (Pipe Inclined Fixed)—The test pipe shall be inclined at 45 deg with the horizontal. The pipe is not rotated during welding. See Fig. 5.8.1.2(D).

(5) Position 6GR (Test for complete joint penetration groove welds of tubular T-, K-, and Y- connections)—The test pipe shall be inclined at 45 deg with the horizontal. The pipe or tube is not rotated during welding. See Fig. 5.8.1.2(E).

5.8.1.3 Fillet Welds (illustrated in Fig. 5.8.1.3). In making the tests to qualify fillet welds, test plates shall be welded in the positions outlined below:

(1) Position 1F (Flat)—The test plates shall be so placed that each fillet weld is deposited with its axis approximately horizontal and its throat approximately vertical. See Fig. 5.8.1.3(A).

(2) Position 2F (Horizontal)—The test plates shall be so placed that each fillet weld is deposited on the upper side of the horizontal surface and against the vertical surface. See Fig. 5.8.1.3(B).

(3) Position 3F (Vertical)—The test plates shall be placed in approximately vertical planes and each fillet

weld deposited on the vertical surfaces. See Fig. 5.8.1.3(C).

(4) Position 4F (Overhead)—The test plates shall be so placed that each fillet weld is deposited on the under side of the horizontal surface and against the vertical surface. See Fig. 5.8.1.3(D).

5.9 Joint Welding Procedure

5.9.1 The joint welding procedure shall comply in all respects with the procedure specification.

5.9.2 Weld cleaning shall be done with the test weld in the same position as the welding position being qualified.

5.10 Test Specimens: Number, Type, and Preparation

5.10.1 Complete Joint Penetration Groove Welds

5.10.1.1 The type and number of specimens that must be tested to qualify a welding procedure are shown in

Table 5.10.1
Number and type of test specimens and range of thickness qualified—procedure qualification; complete joint penetration groove welds

1. Tests on plate

Plate thickness tested, in.	Number of sample welds per position	NDT*	Reduced section tension (see Fig. 5.10.1.3F)	Root bend (see Fig. 5.10.1.3J)	Face bend (see Fig. 5.10.1.3J)	Side bend (see Fig. 5.10.1.3H)	Plate thickness qualified, maximum in.
			Test specimens required				
3/8	1	Yes	2	2	2	—	3/4
Under 3/4 (excluding 3/8)	1	Yes	2	—	—	4	Thickness tested
Over 3/4 and under 1	1	Yes	2	—	—	4	Thickness tested
1 and over	1	Yes	2	—	—	4	Unlimited

Note: All welded test plates shall be visually inspected (see 5.12.7).

*A minimum of 6 in. of effective weld length shall be tested by radiographic or ultrasonic testing prior to mechanical testing (see 5.10.1.3).

2. Tests on pipe or tubing

Pipe size of sample weld — Diam	Wall thickness, t	Number of sample welds per position	NDT**	Reduced section tension (see Fig. 5.10.1.3F)	Root bend (see Fig. 5.10.1.3J)	Face bend (see Fig. 5.10.1.3J)	Side bend (see Fig. 5.10.1.3H)	Pipe or tube size qualified — Diameter, in.	Wall thickness, in. min	Wall thickness, in. max
2 in. or 3 in.	Sch. 80 / Sch. 40	2	Yes	2	2	2	—	3/4 through 4	0.063	0.674
6 in. or 8 in.	Sch. 80 / Sch. 120	1	Yes	2	—	—	4	4 and over	0.187	Any

Job size pipe or tubing

Diam	Wall thickness, t							Diameter, in.	min	max
<24 in.	t<3/4 in.	1	Yes	2	—	—	4	Test diam and over	t/2	2t
	t≥3/4 in.	1	Yes	2	—	—	4	Test diam and over	0.375	Any
≥24 in.	t<3/4 in.	1	Yes	2	—	—	4	24 and over	t/2	2t
	t≥3/4 in.	1	Yes	2	—	—	4	24 and over	0.375	Any

Note: All welded test pipes or tubing shall be visually inspected (see 5.12.6).
**For pipe or tubing, the full circumference of the completed weld shall be tested by RT or UT prior to mechanical testing (5.10.1.3).

Table 5.10.1 (continued)
Number and type of test specimens and range of thickness qualified—procedure qualification;
complete joint penetration groove welds

3. Tests on electroslag and electrogas welding

Plate thickness tested	Number of sample welds	NDT**	Reduced section tension (see Fig. 5.10.1.3F)	All-weld-metal tension (see Fig. 5.10.1.3G)	Side bend (see Fig. 5.10.1.3H)	Impact tests*** (see 4.15.3)	Plate thickness qualified
				Test specimens required			
T*	1	Yes	2	1	4	5	0.5T-1.1T

Note: All welded test plates shall be visually inspected (see 5.12.7).

*T is the test plate thickness.
**6 in. minimum length of weld shall be tested by radiographic or ultrasonic methods prior to mechanical testing (see 5.10.1.3).
***If specified.

Table 5.10.1, together with the range of thickness that is qualified for use in construction. The range is based on the thickness of the test plate, pipe, or tubing used in making the qualification.

5.10.1.2 Test specimens for groove welds in corner or T-joints shall be butt joints having the same groove configuration as the corner or T-joint to be used on construction.

5.10.1.3 Nondestructive Testing. Before preparing mechanical test specimens, the qualification test plate, pipe, or tubing shall be nondestructively tested for soundness as follows:

(1) Either radiographic or ultrasonic testing shall be used. Test plates, on the portion of the weld between the discard strips, as indicated in Figs. 5.10.1.3C through 5.10.1.3E, shall be tested, except that a minimum of 6 in. (152 mm) of effective weld length shall be tested. For pipe or tubing the full circumference of the completed weld shall be tested.

(2) For acceptable qualification the weld, as revealed by radiographic or ultrasonic testing, shall conform to the requirements of 5.12.1.5.

5.10.1.4 Mechanical Testing. The welded test assemblies conforming to 5.10.1.3 shall have test specimens prepared by cutting the test plate, pipe, or, tubing as shown in Figs. 5.10.1.3A through 5.10.1.3E, whichever is applicable. The test specimens shall be prepared for testing in accordance with Figs. 5.10.1.3F through 5.10.1.3J, as applicable.

5.10.2 Partial Joint Penetration Groove Welds. The type and number of specimens that must be tested to qualify a welding procedure are shown in Table 5.10.2. A sample weld shall be made using the type of groove design and joint welding procedure to be used in construction, except the depth of groove need not exceed 1 in.

(25.4 mm). If the partial joint penetration groove weld is to be used for corner or T-joints, the butt joint shall have a temporary restrictive plate in the plane of the square face to simulate the T-joint configuration. The sample welds shall be tested as follows.

5.10.2.1 For joint welding procedures which conform in all respects to Sections 3 and 4, three macroetch cross section specimens shall be prepared to demonstrate that the designated effective throat (obtained from the requirements of the procedure specification) are met.

5.10.2.2 When a joint welding procedure has been qualified for a complete joint penetration groove weld and is applied to the welding conditions of a partial joint penetration groove weld, three macroetch cross section test specimens are required.

5.10.2.3 If a joint welding procedure is not covered by either 5.10.2.1 or 5.10.2.2, or if the welding conditions do not meet a prequalified status, or if they have not been used and tested for a complete joint penetration butt weld, then a sample joint must be prepared and the first operation is to make a macroetch test specimen to determine the effective throat of the joint. Then, the excess material is machined off, on the bottom side of the joint, to the thickness of the effective throat. Tension and bend test specimens shall be prepared and tests performed, as required for complete joint penetration groove welds (see 5.10.1).

5.10.3 Fillet Welds. The type and number of specimens that must be tested to qualify a welding procedure are shown in Table 5.10.3.

5.10.3.1 Fillet Welds. A T-test fillet weld, as shown in Fig. 5.10.3, shall be made for each procedure and position to be used in construction. One test weld shall be the maximum size single pass fillet weld and one test weld shall be the minimum size multiple pass fillet weld used

Table 5.10.2
Number and type of test specimens and range of thickness qualified—procedure qualification;
partial joint penetration groove welds

			Test specimens required			
				Tension and bend tests (5.10.2.3)		
Groove type	Groove depth, max	Number of sample welds	Macroetch for effective throat (E) (5.10.2.1) (5.10.2.2) (5.10.2.3)	Reduced section tension (see Fig. 5.10.1.3F)	Side bend (see Fig. 5.10.1.3H)	Plate thickness qualified, max
Same as used in construction*	1 in.	1	3	2	4	Unlimited

Note: All welded test plates shall be visually inspected (see 5.12.7).

* If a partial joint penetration bevel- or J-groove weld is to be used for T-joints or a double-bevel- or double-J-groove weld is to be used for corner joints, the butt joint shall have a temporary restrictive plate in the plane of the square face to simulate a T-joint configuration.

Table 5.10.3
Number and type of test specimens and range of thickness qualified—procedure qualification;
fillet welds

			Test specimens required			Sizes qualified	
Test specimen	Fillet size	Number of welds per procedure	Macroetch 5.10.3 5.11.2	All-weld-metal tension* (see Fig. 5.10.1.3G)	Side bend (see Fig. 5.10.1.3H)	Plate thickness	Fillet size
T-test (Fig. 5.10.3)	Single pass, max size to be used in construction	1 in each position to be used	3 faces	—	—	Unlimited	Max tested single pass and smaller
	Multiple pass, min size to be used in construction	1 in each position to be used	3 faces	—	—	Unlimited	Min tested multiple pass and larger
Groove test* (Fig. 5.10.3.2 with steel backing)	—	1 in 1G position	—	1	2	Qualifies welding consumables to be used in T-test above	

Note: All welded test plates shall be visually inspected (see 5.12.7).

* When the welding consumables used do not conform to the prequalified provisions of 5.1.1, and a welding procedure using the proposed welding consumables has not been established by the contractor in accordance with either 5.10.1 or 5.10.2, a complete joint penetration groove weld test plate shall be welded in accordance with 5.10.1 (see Fig. 5.10.1.3H)

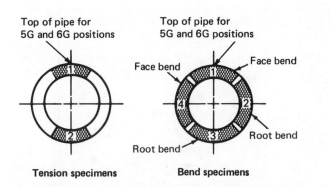

Tension specimens   Bend specimens

Fig. 5.10.1.3A—Location of test specimens on welded test pipe—2 in. or 3 in. in diameter

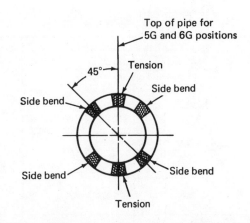

Fig. 5.10.1.3B—Location of test specimens for welded test pipe—6 in. or 8 in. in diameter

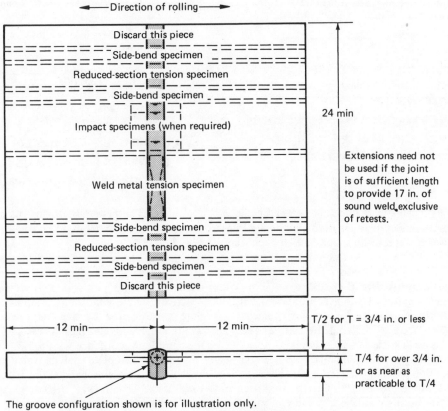

The groove configuration shown is for illustration only.
The groove shape used shall conform to that being qualified.

Fig. 5.10.1.3C—Location of test specimens on welded test plate—electroslag and electrogas welding—procedure qualification

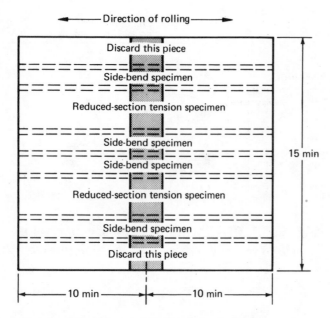

When impact tests are required, the specimens shall be removed from their locations, as shown in Fig. 5.10.1.3C.

The groove configuration shown is for illustration only.
The groove shape used shall conform to that being qualified.

Fig. 5.10.1.3D—Location of test specimens on welded test plate over 3/4 in. thick— procedure qualification

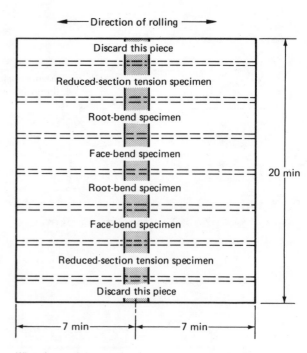

When impact tests are required, the specimens shall be removed from their locations, as shown in Fig. 5.10.1.3C.

The groove configuration shown is for illustration only.
The groove shape used shall conform to that being qualified.

Fig. 5.10.1.3E—Location of test specimens on welded test plate 3/8 in. thick— procedure qualification

in construction. The weldment shall be cut perpendicular to the direction of welding at three locations, as shown in Fig. 5.10.3. Specimens representing one face of each of three cuts shall constitute a macroetch test specimen and shall be tested in accordance with 5.11.2.

5.10.3.2 Consumables Verification Test

(1) If both the proposed welding consumables and the proposed welding procedures for welding the fillet weld test plate prescribed in 5.10.3.1 are neither prequalified nor otherwise qualified by 5.2, that is:

(a) If the welding consumables used do not conform to the prequalified provisions of 5.1.1, and also

(b) If the welding procedure using the proposed consumables has not been established by the contractor in accordance with either 5.10.1 or 5.10.2,

then a complete joint penetration groove weld test plate shall be welded to qualify the proposed combination.

(2) The test plate shall be welded as follows:

(a) The test plate shall have the groove configuration shown in Fig. 5.18A (Fig. 5.34.1 for SAW), with steel backing.

(b) The plate shall be welded in the 1G (flat) position.

(c) The plate length shall be adequate to provide the test specimens required below, oriented as shown in Fig. 5.10.3.2.

(d) The welding test conditions of current, voltage, travel speed, and gas flow shall approximate those to be used in making production fillet welds as closely as practical. These conditions establish the welding procedure specification from which, when production fillet welds are made, changes in essential variables will be measured in accordance with 5.5.2.

(3) The test plate shall be tested as follows:

(a) Two side-bend (Fig. 5.10.1.3H) and one all-weld-metal tension (Fig. 5.10.1.3G) test specimens shall be removed from the test plate, as shown in Fig. 5.10.3.2.

(b) The bend test specimens shall be tested in accordance with 5.11.3. Those test results shall conform to the requirements of 5.12.1.2.

(c) The tension test specimen shall be tested in accordance with 5.11.4. The test result shall determine the strength level of the welding consumables, which shall conform to the requirements of 4.1.2 for the welding process being used and the base metal strength level being welded.

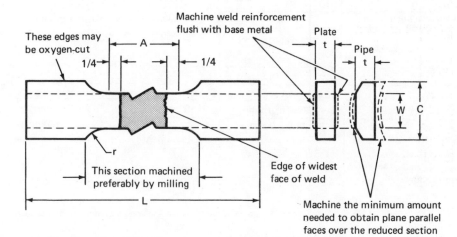

	Dimensions				
	Test plate			Test pipe	
				2 in. & 3 in. diameter	6 in. & 8 in. diameter or large job size pipe
	Tp ≤ 1 in.	1 < Tp < 1-1/2 in.	Tp ≥1-1/2 in.		
A—Length of reduced section	Widest face of weld + 1/2 in., 2-1/4 min			Widest face of weld + 1/2 in., 2-1/4 min	
L—Overall length, min (Note 2)	As required by testing equipment			As required by testing equipment	
W—Width of reduced section (Notes 3,4)	1-1/2 ± 0.01	1 ± 0.01	1 ± 0.01	1/2 ±0.01	3/4 ± 0.01
C—Width of grip section, min (Notes 4,5)	2	1-1/2	1-1/2	1 approx.	1-1/4 approx.
t —Specimen thickness (Notes 6,7)	Tp	Tp	Tp/n (Note 7)	Maximum possible with plane parallel faces within length A	
r —Radius of fillet, min	1/2	1/2	1/2	1	1

Notes:

1. Tp = thickness of the plate.
2. It is desirable, if possible, to make the length of the grip section large enough to allow the specimen to extend into the grips a distance equal to two-thirds or more of the length of the grips.
3. The ends of the reduced section shall not differ in width by more than 0.004 in. Also, there may be a gradual decrease in width from the ends to the center, but the width at either end shall not be more than 0.015 in. larger than the width at the center.
4. Narrower widths (W and C) may be used when necessary. In such cases, the width of the reduced section should be as large as the width of the material being tested permits. If the width of the material is less than W, the sides may be parallel throughout the length of the specimen.
5. For standard plate-type specimens, the ends of the specimen shall be symmetrical with the center line of the reduced section within 0.25 in. except for referee testing, in which case the ends of the specimen shall be symmetrical with the center line of the reduced section within 0.10 in.
6. The dimension t is the thickness of the test specimen as provided for in the applicable material specifications. The minimum nominal thickness of 1-1/2 in. wide specimens shall be 3/16 in. except as permitted by the product specification.
7. For plates over 1-1/2 in. thick, specimens may be cut into the minimum number of approximately equal strips not exceeding 1-1/2 in. in thickness. Test each strip and average the results.

Fig. 5.10.1.3F—Reduced-section tension specimens

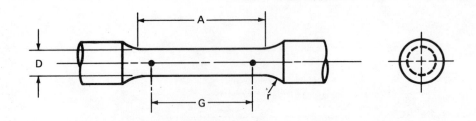

Dimensions			
	Standard specimen	Small-size specimens proportional to standard	
Nominal diameter	0.500 in. round	0.350 in. round	0.250 in. round
G — Gage length	2.000 ± 0.005	1.400 ± 0.005	1.000 ± 0.005
D — Diameter (Note 1)	0.500 ± 0.010	0.350 ± 0.007	0.250 ± 0.005
r — Radius of fillet, min	3/8	1/4	3/16
A — Length of reduced section (Note 2), min	2-1/4	1-3/4	1-1/4

Notes:

1. The reduced section may have a gradual taper from the ends toward the center, with the ends not more than one percent larger in diameter than the center (controlling dimension).
2. If desired, the length of the reduced section may be increased to accommodate an extensometer of any convenient gage length. Reference marks for the measurement of elongation should be spaced at the indicated gage length.
3. The gage length and fillets shall be as shown, but the ends may be of any form to fit the holders of the testing machine in such a way that the load shall be axial. If the ends are to be held in wedge grips, it is desirable, if possible, to make the length of the grip section great enough to allow the specimen to extend into the grips a distance equal to two-thirds or more of the length of the grips.

Fig. 5.10.1.3G — All-weld-metal tension specimens

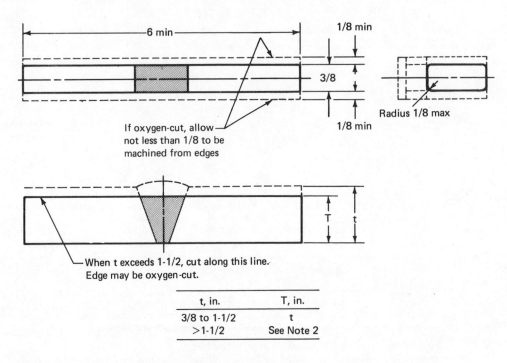

t, in.	T, in.
3/8 to 1-1/2	t
>1-1/2	See Note 2

Notes:

1. A longer specimen length may be necessary when using a wraparound-type bending fixture or when testing steel with a yield point of 90 ksi or more.
2. For plates over 1-1/2 in. thick, cut the specimen into approximately equal strips with T between 3/4 and 1-1/2 in. and test each strip.
3. t = plate or pipe thickness.

Fig. 5.10.1.3H — Side-bend specimens

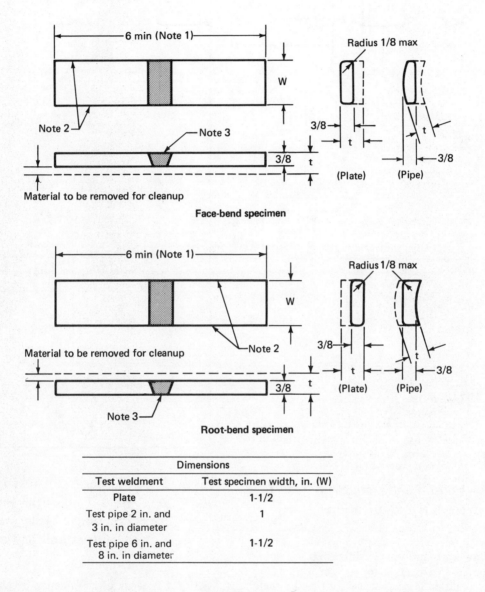

Dimensions	
Test weldment	Test specimen width, in. (W)
Plate	1-1/2
Test pipe 2 in. and 3 in. in diameter	1
Test pipe 6 in. and 8 in. in diameter	1-1/2

Notes:

1. A longer specimen length may be necessary when using a wraparound-type bending fixture or when testing steel with a yield point of 90 ksi or more.

2. These edges may be oxygen-cut and may or may not be machined.

3. The weld reinforcement and backing, if any, shall be removed flush with the surface of the specimen (see 3.6.3). If a recessed backing is used, this surface may be machined to a depth not exceeding the depth of the recess to remove the backing; in such cases, the thickness of the finished specimen shall be that specified above. Cut surfaces shall be smooth and parallel.

4. t = plate or pipe thickness.

Fig. 5.10.1.3J — Face- and root-bend specimens

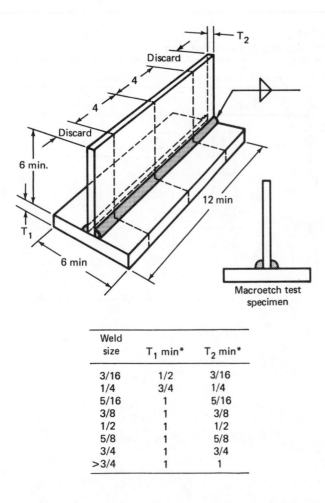

Weld size	T_1 min*	T_2 min*
3/16	1/2	3/16
1/4	3/4	1/4
5/16	1	5/16
3/8	1	3/8
1/2	1	1/2
5/8	1	5/8
3/4	1	3/4
>3/4	1	1

*Note: Where the maximum plate thickness used in production is less than the value shown in the table, the maximum thickness of the production pieces may be substituted for T_1 and T_2.

Fig. 5.10.3—Fillet weld soundness test for procedure qualification

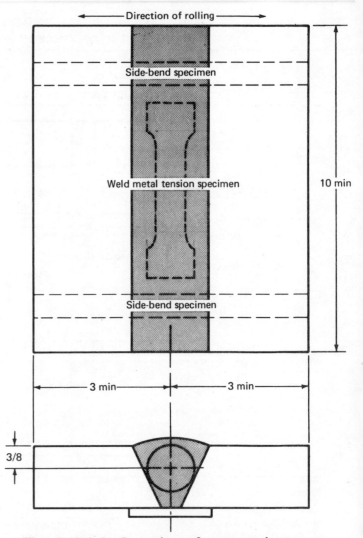

Fig. 5.10.3.2—Location of test specimens on welded test plate 1 in. thick—consumables verification for fillet weld procedure qualification

5.10.3.3 Pipe and Tubing Qualification. A joint welding procedure specification for groove welding of pipe or tubing qualified in accordance with 5.10.1 shall also constitute procedure qualification for fillet welding pipe or tubing in the same position qualified.

5.10.4 Test Plate Aging. When required by the filler metal specification applicable to weld metal being tested, fully welded qualification test plates may be aged at 200 to 220° F (93 to 104° C) for 48 ± 2 hours.

5.10.5 Pipe Welding Positions Qualified. Qualification on pipe shall also qualify for plate, but not vice versa, except that qualification on plate in the 1G (flat) or 2G (horizontal) positions shall qualify for welding pipe or tubing over 24 in. (600 mm) in diameter. Welding position limitations for procedure qualification are shown in Table 5.10.5.

5.10.5.1 Procedure qualification of pipe or tubing in the 5G (pipe horizontal fixed) position qualifies the procedure for flat, vertical, and overhead position groove welding of pipe, tubing, and plate.

5.10.5.2 Procedure qualification of pipe or tubing in the 6G (inclined fixed) position qualifies the procedure for all position groove welding of pipe, tubing, and plate, but does not qualify the procedure for groove welding of T-, Y-, and K-connections.

5.10.5.3 Procedure qualification of T-, Y-, and K-connections in accordance with 10.13.1.1 qualifies for groove T-, Y-, and K-connections, subject to the limitations of 10.13.1.1. It also qualifies for groove and fillet welding in all positions of pipe and tubing, and groove welding in all positions of plate.

Table 5.10.5
Procedure qualification—type and position limitations

| Qualification test | | Type of weld and position of welding qualified* | | | |
| | | Plate | | Pipe | |
Weld	Plate or pipe positions**	Groove	Fillet	Groove	Fillet
Plate—groove Complete joint penetration	1G 2G 3G 4G	F H V OH		F (Note 1) H (Note 1)	
Plate—groove Partial joint penetration	1G 2G 3G 4G	F H V OH		F (Note 1) H (Note 1)	
Plate—fillet	1F 2F 3F 4F		F H V OH		
Pipe—groove Complete joint penetration	1G 2G 5G 6G T-, Y-, K-connections	F F, H F,V,OH F,H,V,OH (Note 2) All		F F, H F,V,OH F,H,V,OH (Note 2) All (Note 3)	F F, H F,V,OH F,H,V,OH (Note 2) All

Notes:
1. Qualifies for welding pipe or tubing over 24 in. (610 mm) in diameter.
2. Qualifies for complete joint penetration groove welds in all positions except T-, Y-, and K- connections.
3. Qualifies for T-, Y-, and K-connections subject to limitations of 10.13.1.1, and for complete joint penetration groove welds in all positions.

*Positions of welding: F = flat, H = horizontal, V = vertical, OH = overhead.
**See Figs. 5.8.1, 5.8.1.2, and 5.8.1.3.

5.11 Method of Testing Specimens

5.11.1 Reduced-Section Tension Specimens. Before testing, the least width and corresponding thickness of the reduced section shall be measured in inches. The specimen shall be ruptured under tensile load, and the maximum load in pounds shall be determined. The cross-sectional area shall be obtained by multiplying the width by the thickness. The tensile strength in psi shall be obtained by dividing the maximum load by the cross-sectional area.

5.11.2 Macroetch Test. The weld test specimens shall be prepared with a finish suitable for macroetch examination. A suitable solution shall be used for etching to give a clear definition of the weld.

5.11.3 Root-, Face-, and Side-Bend Specimens. Each specimen shall be bent in a jig having the contour shown in Fig. 5.27.1 and otherwise substantially in accordance with that figure. Any convenient means may be used to move the plunger member with relation to the die member.

The specimen shall be placed on the die member of the jig with the weld at midspan. Face-bend specimens shall be placed with the face of the weld directed toward the gap. Root-bend and fillet-weld-soundness specimens shall be placed with the root of the weld directed toward the gap. Side-bend specimens shall be placed with that side showing the greater discontinuity, if any, directed toward the gap.

The plunger shall force the specimen into the die until the specimen becomes U-shaped. The weld and heat-affected zones shall be centered and completely within the bent portion of the specimen after testing.

When using the wraparound jig, the specimen shall be firmly clamped on one end so that there is no sliding of the specimen during the bending operation. The weld and heat-affected zones shall be completely in the bent portion of the specimen after testing. Test specimens shall be removed from the jig when the outer roll has been moved 180 deg from the starting point.

5.11.4 All-Weld-Metal Tension Test. The test specimen shall be tested in accordance with ASTM A370, Mechanical Testing of Steel Products.

5.11.5 The radiographic procedure and technique shall be in accordance with the requirements of Part B of Section 6 of this code.

5.11.6 The ultrasonic procedure and technique shall be in accordance with the requirements of Part C of Section 6 of this code.

5.12 Test Results Required

The requirements for the test results shall be as follows.

5.12.1 Reduced-Section Tension Test. The tensile strength shall be no less than the minimum of the specified tensile range of the base metal used.

5.12.2 Root-, Face-, and Side-Bend Tests. The convex surface of the specimen shall be examined for the appearance of cracks or other open discontinuities. Any specimen in which a crack or other open discontinuity exceeding 1/8 in. (3.2 mm) measured in any direction is present after the bending shall be considered as having failed. Cracks occurring on the corners of the specimen during testing shall not be considered.

5.12.3 Macroetch Tests. The specimens shall be examined for discontinuities and any which have discontinuities prohibited by 8.15, 9.25, or 10.17, whichever is applicable, shall be considered as failed. The specimens shall have the designated effective throat for partial joint penetration groove welds. Fillet weld specimens shall show fusion to the root but not necessarily beyond, and both legs shall be equal to within 1/8 in. (3.2 mm). Convexity shall not exceed the limits specified in 3.6.1.

5.12.4 All-Weld-Metal Tension Test (electroslag and electrogas). The mechanical properties shall be no less than those specified in Table 4.16.

5.12.5 Nondestructive Testing. For acceptable qualification, the weld, as revealed by radiographic or ultrasonic testing, shall conform to the requirements of 8.15, 9.25, or 10.17, whichever is applicable.

5.12.6 Visual Inspection—Pipe and Tubing. For acceptable qualification, a pipe weld, when inspected visually, shall conform to the following requirements:

(1) The weld shall be free of cracks.

(2) All craters shall be filled to the full cross section of the weld.

(3) The face of the weld shall be at least flush with the outside surface of the pipe, and the weld shall merge smoothly with the base metal. Undercut shall not exceed 1/64 in. (0.4 mm). Weld reinforcement shall not exceed the following:

Pipe wall thickness	Reinforcement, max	
in. (mm)	in.	mm
3/8 (9.5) or less	3/32	2.4
Over 3/8 to 3/4 (19.0) incl.	1/8	3.2
Over 3/4	3/16	4.8

(4) The root of the weld shall be inspected, and there shall be no evidence of cracks, incomplete fusion, or inadequate joint penetration. A concave root surface is permitted within the limits shown below, provided the total weld thickness is equal to or greater than that of the base metal.

(5) The maximum root surface concavity shall be 1/16 in. (1.6 mm) and the maximum melt-thru shall be 1/8 in. (3.2 mm).

5.12.7 Visual Inspection—Plate. For acceptable qualification, the welded test plate, when inspected visually, shall conform to the requirements for visual inspection in 9.25.1.

5.13 Records

Records of the test results shall be kept by the manufacturer or contractor and shall be available to those authorized to examine them.

5.14 Retests

If any one specimen of all those tested fails to meet the test requirements, two retests for that particular type of test specimen may be performed with specimens cut from the same procedure qualification test material. The results of both retest specimens must be the test requirements. For material over 1-1/2 in. (38.1 mm) thick, failure of a specimen shall require testing of all specimens of the same type from two additional locations in the test material.

Part C
Welder Qualification

5.15 General

The qualification tests described in Part C are specially devised tests to determine the welder's ability to produce sound welds. The qualification tests are not intended to be used as a guide for welding during actual construction. The latter shall be performed in accordance with the requirements of the procedure specification.

5.16 Limitation of Variables

For the qualification of a welder the following rules shall apply:

5.16.1 Qualification established with any one of the steels permitted by this code shall be considered as qualification to weld or tack weld any of the other steels.

5.16.2 A welder shall be qualified for each process used.

5.16.3 A welder qualified for shielded metal arc welding with an electrode identified in the following table shall be considered qualified to weld or tack weld with any other electrode in the same group designation and with any electrode listed in a numerically lower group designation.

Group designation	AWS electrode classification*
F4	EXX15, EXX16, EXX18
F3	EXX10, EXX11
F2	EXX12, EXX13, EXX14
F1	EXX20, EXX24, EXX27, EXX28

*The letters "XX" used in the classification-designation in this table stand for the various strength levels (60, 70, 80, 90, 100, and 120) of deposited weld metal.

5.16.4 A welder qualified with an approved electrode and shielding medium combination shall be considered qualified to weld or tack weld with any other approved electrode and shielding medium combination for the process used in the qualification test.

5.16.5 A change in the position of welding to one for which the welder is not already qualified shall require requalification.

5.16.6 A change from one diameter-wall pipe grouping shown in Table 5.26.1 to another shall require requalification.

5.16.7 When the plate is in the vertical position, or the pipe or tubing is in the 5G or 6G position, a change in the direction of welding shall require requalification.

5.16.8 The omission of backing material in complete joint penetration welds welded from one side shall require requalification.

5.17 Qualification Tests Required

5.17.1 The welder qualification tests for manual and semiautomatic welding shall be as follows:

5.17.1.1 Groove weld qualification test for plate of unlimited thickness.

5.17.1.2 Groove weld qualification test for plate of limited thickness.

5.17.1.3 Fillet weld qualification tests for fillet welds only.

(1) For welds in joints having a dihedral angle (ψ) of 75 deg or less, qualification tests shall be as required by 5.18 or 5.19. Such qualification will be valid for fillet welds having angles greater than 75 deg.

(2) For welds in joints having a dihedral angle(ψ) greater than 75 deg and not exceeding 135 deg, tests shall be as required by 5.22, Option 1 or Option 2—contractor's option.

5.17.2 The pipe or tubing qualification tests for manual and semiautomatic welding shall be as follows:

5.17.2.1 Groove weld qualification test for butt joints on pipe or square or rectangular tubing.

5.17.2.2 Groove weld qualification test for T-, K-, or Y-connections on pipe or square or rectangular tubing.

5.17.2.3 Groove weld qualification test for butt joints on square or rectangular tubing tested on flat plate.

5.17.3 The welder who makes a complete joint penetration plate groove weld procedure qualification test that meets the requirements is thereby qualified for that process and test position for plates and square or rectangular tubing equal to or less than the thickness of the test plate welded. If the test plate is 1 in. (25.4 mm) or greater in thickness, the welder will be qualified for all thicknesses. The welder is also qualified for fillet welding of plate and pipe, as shown in Table 5.23.

5.17.4 The welder who makes a complete joint penetration groove weld pipe procedure qualification test, without backing strip, that meets the requirements is thereby qualified for that process. His qualification will include the test position for pipe having a wall thickness equal to or less than the wall thickness of the test pipe welded. If the test pipe welded is 6 in. (152 mm) Sch. 80 or 8 in. (203 mm) Sch. 120 pipe, he will be qualified for all thicknesses. This welder is also qualified for fillet welding of plate and pipe as shown in Table 5.23. If the diameter of the job-size pipe or tubing used in qualification is 4 in. (102 mm) or less, the qualification is limited

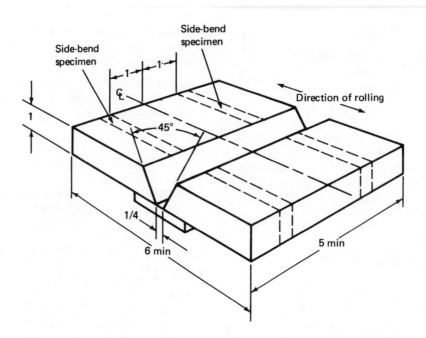

Note: When radiography is used for testing, no tack welds shall be in the test area.

Fig. 5.18A—Test plate for unlimited thickness—welder qualification

to diameters 3/4 in. (19 mm) through 4 in.. (102 mm), inclusive. If the diameter of job-size pipe is over 4 in. (102 mm), the qualification is limited to a minimum diameter of greater than 1/2 test diameter or 4 in. (102 mm), whichever is larger. The wall thickness qualified and the number of test specimens required shall be the same as specified for the equivalent pipe size in Table 5.26.1.

5.18 Groove Weld Plate Qualification Test for Plate of Unlimited Thickness

The joint detail shall be as follows: 1 in. (25.4 mm) plate, single-V-groove, 45 deg included angle, 1/4 in. (6.4 mm) root opening with backing (see Fig. 5.18A). For horizontal position qualification, the joint detail may, at the contractor's option, be as follows: single-bevel-groove, 45 deg groove angle, 1/4 in. root opening with backing (see Fig. 5.18B). Backing must be at least 3/8 in. (9.5 mm) by 3 in. (76.2 mm) if radiographic testing is used without removal of backing. It must be at least 3/8 in. by 1 in. (25.4 mm) for mechanical testing or for radiographic testing after the backing is removed. Minimum length of welding groove shall be 5 in. (127 mm).

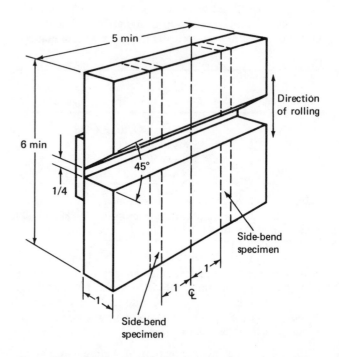

Note: When radiography is used for testing, no tack welds shall be in the test area.

Fig. 5.18B—Optional test plate for unlimited thickness—horizontal position—welder qualification

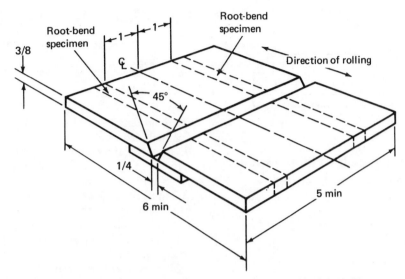

Note: When radiography is used for testing, no tack welds shall be in the test area.

Fig. 5.19A—Test plate for limited thickness—all positions—welder qualification

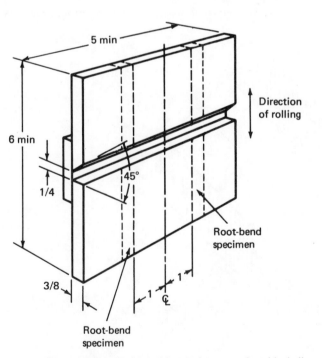

Note: When radiography is used for testing, no tack welds shall be in the test area.

Fig. 5.19B—Optional test plate for limited thickness—horizontal position—welder qualification

5.19 Groove Weld Plate Qualification Test for Plate of Limited Thickness

The joint detail shall be as follows: 3/8 in. (9.5 mm) plate, single-V-groove, 45 deg included angle, 1/4 in. (6.4 mm) root opening with backing (see Fig. 5.19A). For horizontal position qualification, the joint detail may, at the contractor's option, be as follows: single-bevel-groove, 45 deg groove angle, 1/4 in. root opening with backing (see Fig. 5.19B). Backing must be at least 3/8 in. by 3 in. (76.2 mm) if radiographic testing is used without removal of backing. It must be at least 3/8 in. by 1 in. (25.4 mm) for mechanical testing or for radiographic testing after the backing is removed. Minimum length of welding groove shall be 5 in. (127 mm).

5.20 Groove Weld Qualification Test for Butt Joints on Pipe or Square or Rectangular Tubing

The joint detail shall be that shown in a qualified joint welding procedure specification for a single-welded pipe butt weld or shall be as follows: pipe diameter-wall thickness as required, single-V-groove, 60 deg included angle, 1/8 in. (3.2 mm) max root face and root opening without backing strip (see Fig. 5.20A), or single-V-groove, 60 deg included angle and suitable root opening with backing (see Fig. 5.20B).

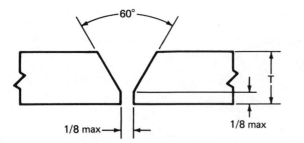

Fig. 5.20A—Plate and pipe butt joint—welder qualification—without backing

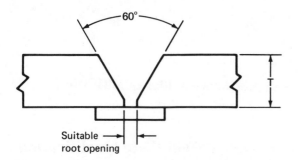

Fig. 5.20B—Pipe test butt joint—welder qualification—with backing

5.21 Groove Weld Qualification Test for T-, K-, or Y-Connections on Pipe or Square or Rectangular Tubing

The joint detail shall be as follows: single-bevel, 37-1/2 deg included angle with bevel on pipe or tube at least 1/2 in. thick; the square edge pipe or tube shall be at least 3/16 in. (4.8 mm) thicker than the beveled pipe thickness; 1/16 in. (1.6 mm) max root race and 1/8 in. (3.2 mm) root opening. A restriction ring shall be placed on the thicker material, within 1/2 in. (12.7 mm) of the joint and shall extend at least 6 in. (152 mm) beyond the surface of the pipe or tube. (See Fig. 5.21). Test specimens for side bends shall be taken as indicated in Fig. 5.26.2 and machined to be standard specimens with parallel sides.

5.22 Fillet Weld Qualification Test for Fillet Welds Only

For fillet weld qualification only: (1) For fillet welds between parts having a dihedral angle (ψ) of 75 deg or less, the welder shall weld a groove weld test plate as

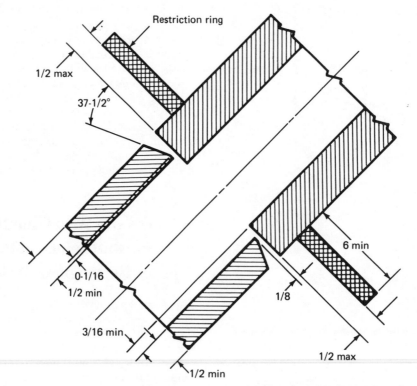

Fig. 5.21—Test joint for T-, K-, and Y-connections on pipe or square or rectangular tubing—welder qualification

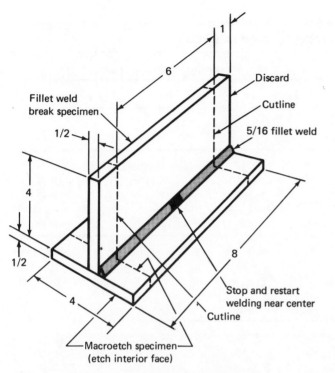

Note: Plate thickness and dimensions are minimum.

Fig. 5.22.1—Fillet-weld-break and macroetch test plate—welder qualification—option 1

required by 5.18 or 5.19; and (2) for joints having a dihedral angle (ψ) greater than 75 deg, but not exceeding 135 deg, the welder shall weld a test plate according to Option 1 or Option 2 depending on the contractor's choice, as follows:

5.22.1 Option 1. Weld a T-test plate in accordance with Fig. 5.22.1.

5.22.2 Option 2. Weld a soundness test plate in accordance with Fig. 5.22.2.

5.23 Position of Test Welds

(See Table 5.23.)

5.23.1 Groove Plate Test Welds

5.23.1.1 Qualification in the 1G (flat) position qualifies for flat position groove welding of plate, pipe, and tubing; flat and horizontal position fillet welding of plate; and flat and horizontal position fillet welding of pipe and tubing.

5.23.1.2 Qualification in the 2G (horizontal) position qualifies for flat and horizontal position groove and flat and horizontal position fillet welding of plate, pipe, and tubing.

5.23.1.3 Qualification in the 3G (vertical) position qualifies for flat, horizontal, and vertical position groove

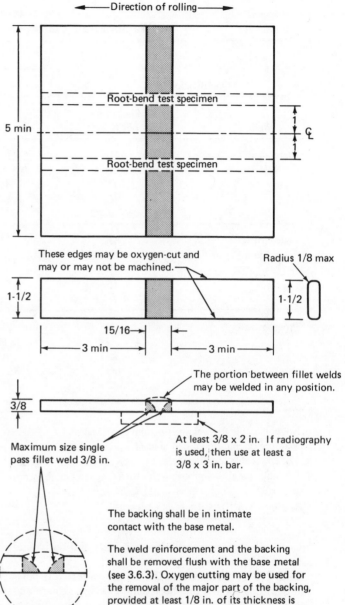

Fig. 5.22.2—Fillet-weld root-bend test plate —welder qualification—option 2

and flat, horizontal, and vertical position fillet welding of plate; and flat and horizontal position fillet welding of pipe and tubing.

5.23.1.4 Qualification in the 4G (overhead) position qualifies for flat and overhead position groove and flat, horizontal, and overhead position fillet welding of plate; and flat position fillet welding of pipe and tubing.

5.23.2 Groove Pipe Test Welds

5.23.2.1 Qualification in the 1G (pipe horizontal rolled) position qualifies for flat position groove welding of pipe, tubing, and plate; flat and horizontal position

Table 5.23
Welder qualification—type and position limitations

| Qualification test | | Type of weld and position of welding qualified* | | | |
| | | Plate | | Pipe | |
Weld	Plate or pipe positions**	Groove	Fillet	Groove	Fillet
Plate—groove	1G	F	F H	F (Note 2)	F H
(Note 1)	2G	F H	F H	F H (Note 2)	F H
	3G	F H V	F H V		F H
	4G	F OH	F H OH		F
	3G & 4G	All	All		F H
Plate—fillet	1F		F		F
(Notes 1 and 3)	2F		F, H		F, H
	3F		F, H, V		
	4F		F, H, OH		
	3F & 4F		All		
Pipe—groove	1G	F	F H	F	F H
	2G	F H	F H	F H	F H
	5G	F V OH	F V OH	F V OH	F V OH
	6G	Note 4	Note 4	Note 4	Note 4
	2G & 5G	Note 4	Note 4	Note 4	Note 4
	6GR	All	All	All	All

Notes:
1. Not applicable for welding operator qualification (see 5.33.5).
2. Welding operator qualified to weld pipe or tubing over 24 in. in diameter in test position qualified for plate.
3. Not applicable for fillet welds between parts having a dihedral angle of 75 deg or less (see 5.22).
4. Qualifies for all but groove welds for T-, Y-, and K-connections.

*Positions of welding: F=flat, H=horizontal, V=vertical, OH=overhead.
**See Figs. 5.8.1.1, 5.8.1.2, and 5.8.1.3.

fillet welding of pipe and tubing; and flat and horizontal position fillet welding of plate.

5.23.2.2 Qualification in the 2G (pipe vertical) position qualifies for flat and horizontal position groove and flat and horizontal position fillet welding of pipe, tubing, and plate.

5.23.2.3 Qualification in the 5G (pipe horizontal fixed) position qualifies for flat, vertical, and overhead position groove and flat, vertical, and overhead position fillet welding of pipe, tubing, and plate.

5.23.2.4 Qualification in the 6G (inclined fixed) position qualifies for all position groove and all position fillet welding of pipe, tubing, and plate.

5.23.2.5 Qualification for T-, K-, or Y-connections in the 6GR (inclined fixed) position qualifies for groove T-, K-, or Y-connections and groove and fillet welding in all positions of pipe, tubing, and plate.

5.23.3 Fillet Weld Test

5.23.3.1 Qualification in the 1F (flat) position qualifies for flat position fillet welding of plate, pipe, and tubing.

5.23.3.2 Qualification in the 2F (horizontal) position qualifies for flat and horizontal position fillet welding of plate, pipe, and tubing.

5.23.3.3 Qualification in the 3F (vertical) position qualifies for flat, horizontal, and vertical position fillet welding of plate.

5.23.3.4 Qualification in the 4F (overhead) position qualifies for flat, horizontal, and overhead position fillet welding of plate.

5.23.4 Qualification on the groove plate test weld in 1G (flat) or 2G (horizontal) position shall also qualify for butt welding pipe with a backing in the same position qualified. If no backing is used in the groove plate test weld, this shall also qualify for groove welding pipe with or without backing in the same position qualified.

5.24 Base Metal

The base metal used shall comply with 10.2 or the procedure specification.

Table 5.26.1
Number and type of specimens and range of thickness qualified—
welder and welding operator qualification

1. Tests on plate

Type of weld	Thickness of test plate as welded, in.	Visual inspection	Number of specimens						Plate, thickness qualified, in.
			Bend tests* All positions except 5G & 6G			T-joint break	Macro-etch test		
			Face	Root	Side				
Groove**	3/8	Yes	1	1	—	—	—		3/4 max[3]
Groove	Under 1, excluding 3/8	Yes	—	—	2	—	—		Thickness tested and smaller[3]
Groove	1 or over	Yes	—	—	2	—	—		Unlimited[3]
Fillet option No. 1[1]	1/2	Yes	—	—	—	1	1		Unlimited
Fillet option No. 2[2]	3/8	Yes	—	2	—	—	—		Unlimited

1. See Fig. 5.22.1 or 5.34.4.1 as applicable.
2. See Fig. 5.22.2 or 5.34.4.2 as applicable.
3. Also qualifies for welding fillet welds on material of unlimited thickness.

*Radiographic examination of the welder or welding operator test plate may be made in lieu of the bend test. (See 5.3.2.)
**Not applicable for welding operator qualification.

5.25 Joint Welding Procedure

5.25.1 The welder shall follow a joint welding procedure specification applicable to the joint details given in 5.18, 5.19, 5.20, 5.21, or 5.22, whichever is applicable. For complete joint penetration groove welds, welded from one side without backing, the welder shall follow a welding procedure specification applicable to the joint detail shown in Fig. 5.20A.

5.25.2 Weld cleaning shall be done with the test weld in the same position as the welding position being qualified.

5.26 Test Specimens: Number, Type, and Preparation

5.26.1 The type and number of test specimens that must be tested to qualify a welder by mechanical testing are shown in Table 5.26.1 together with the range of thickness that is qualified for use in construction by the thickness of the test plate, pipe, or tubing used in making the qualification. Radiographic testing of the test weld may be used at the contractor's option in lieu of mechanical testing.

5.26.2 Guided-bend test specimens shall be prepared by cutting the test plate, pipe, or tubing as shown in Figs. 5.18A, 5.18B, 5.19A, 5.19B, 5.22.2, or 5.26.2, whichever are applicable, to form specimens approximately rectangular in cross section. The specimens shall be prepared for testing in accordance with Figs. 5.10.1.3F through J, whichever is applicable.

5.26.3 The fillet weld break and macroetch test specimens shall be cut from the test joint, as shown in Fig. 5.22.1. The end of the macroetch test specimen shall be smooth for etching.

5.26.4 If radiographic testing is used in lieu of the prescribed bend tests, the weld reinforcement need not be ground or otherwise smoothed for inspection unless its surface irregularities or juncture with the base metal would cause objectionable weld discontinuities to be obscured in the radiograph. If the backing is removed for radiography, the root shall be ground flush (see 3.6.3) with the base metal.

5.27 Method of Testing Specimens

5.27.1 Root-, Face-, or Side-Bend Specimens. Each specimen shall be bent in a jig having the contour shown

Table 5.26.1 (continued)
Number and type of specimens and range of thickness qualified—
welder and welding operator qualification

2. Tests on pipe or tubing

Type of weld	Pipe or tubing size, as welded		Visual inspection	Number of specimens						Pipe or tube size qualified, in.	Pipe or tube wall thickness qualified, in.	
	Diam	Thickness		All positions except 5G & 6G			5G & 6G positions only				min	max[1]
				Face bend	Root bend	Side bend	Face bend	Root bend	Side bend			
Groove	2 in. or 3 in.	Sch. 80 Sch. 40	Yes	1	1	—	2	2	—	4 or smaller	0.063	0.674[1]
Groove	6 in. or 8 in.	Sch. 80 Sch. 120	Yes	—	—	2	—	—	4	Larger than 4	0.187	Unlimited[1]
Groove	See Fig. 5.21		Yes	—	—	—	—	—	4	T-, K-, and Y- connections		Unlimited[1]

Type of weld	Job size pipe or tubing		Visual inspection	Face bend	Root bend	Side bend	Face bend	Root bend	Side bend	Pipe or tube size qualified, in.	min	max[1]
	Diam	Wall thickness										
Groove	≤4 in.	Any	Yes	1	1	—	2	2	—	3/4 through 4	0.063	0.674[1]
Groove	>4 in.	Any	Yes	—	—	2	—	—	4	1/2 test diam or 4 min[2]	0.187	Unlimited[1]

Note: Radiographic examination of the welder or welding operator test pipe may be made in lieu of the bend test. (See 5.3.2.)

1. Also qualifies for welding fillet welds on material of unlimited thickness.
2. Minimum pipe size qualified shall not be less than 4 in. or 1/2d, whichever is greater, where d is diameter of test pipe.

3. Tests on electroslag and electrogas welds

Plate thickness tested	Number of sample welds	Test specimens required		Plate thickness qualified, in.
		Visual inspection	Side bend (see Fig. 5.10.1.3H)	
1-1/2 max	1	Yes	2	Unlimited for 1-1/2 Max tested for <1-1/2

Note: Radiographic examination of test plate may be made in lieu of the bend test. (See 5.3.2.)

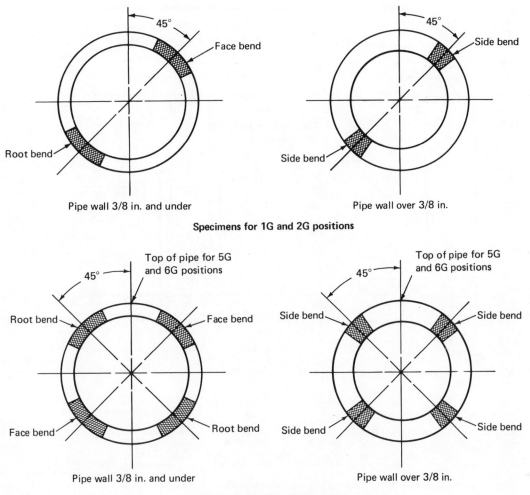

Fig. 5.26.2—Location of test specimens on welded test pipe—welder qualification

in Fig. 5.27.1 and otherwise substantially in accordance with that figure. Any convenient means may be used to move the plunger member with relation to the die member.

The specimen shall be placed on the die member of the jig with the weld at midspan. Face-bend specimens shall be placed with the face of the weld directed toward the gap. Root-bend and fillet weld soundness specimens shall be placed with the root of the weld directed toward the gap. Side-bend specimens shall be placed with that side showing the greater discontinuity, if any, directed toward the gap.

The plunger shall force the specimen into the die until the specimen becomes U-shaped. The weld and heat-affected zones shall be centered and completely within the bent portion of the specimen after testing.

When using a wraparound jig, the specimen shall be firmly clamped on one end so that the specimen does not slide during the bending operation. The weld and heat-affected zones shall be completely within the bent portion

of the specimen after testing. Test specimens shall be removed from the jig when the outer roll has been moved 180 deg from the starting point.

5.27.2 Fillet Weld Break Test. The entire length of the fillet weld shall be examined visually and then the 6 in. (152 mm) long specimen shall be loaded in such a way that the root of the weld is in tension. The load shall be steadily increased or repeated until the specimen fractures or bends flat upon itself.

5.27.3 Macroetch Test. The test specimens shall be prepared with a finish suitable for macroetch examination. A suitable solution shall be used for etching to give a clear definition of the weld.

5.27.4 Radiographic Test. The radiographic procedure and technique shall be in accordance with the requirements of Part B, Section 6. Only the center half of the length of the test plate or 50% of the test pipe shall be subject to testing.

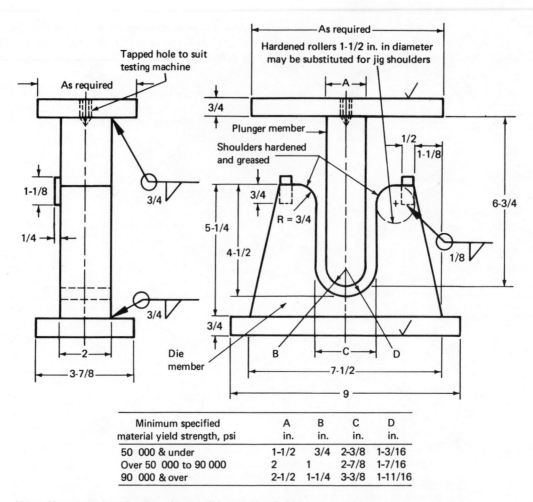

Minimum specified material yield strength, psi	A in.	B in.	C in.	D in.
50 000 & under	1-1/2	3/4	2-3/8	1-3/16
Over 50 000 to 90 000	2	1	2-7/8	1-7/16
90 000 & over	2-1/2	1-1/4	3-3/8	1-11/16

Note: Plunger and interior die surfaces shall be machine-finished.

Fig. 5.27.1A—Guided-bend test jig

5.28 Test Results Required

5.28.1 Root-, Face-, and Side-Bend Tests. The convex surface of the specimen shall be examined for the appearance of cracks or other open discontinuities. Any specimen in which a crack or other open discontinuity exceeding 1/8 in. (3.2 mm) measured in any direction is present after the bending shall be considered as having failed. Cracks occurring on the corners of the specimen during testing shall not be considered.

5.28.2 Fillet Weld Break Test

5.28.2.1 To pass the visual examination, the fillet weld shall present a reasonably uniform appearance and shall be free of overlap, cracks, and excessive undercut. There shall be no porosity visible on the surface of the weld.

5.28.2.2. The specimen shall pass the test if it bends flat upon itself. If the fillet weld fractures, the fractured surface shall show complete fusion to the root of the joint and shall exhibit no inclusion or porosity larger than 3/32 in. (2.4 mm) in greatest dimension. The sum of the greatest dimensions of all inclusions and porosity shall not exceed 3/8 in. (9.5 mm) in the 6 in. (152 mm) long specimen.

5.28.3 Macroetch Test. The specimen shall be examined for discontinuities, and if discontinuities prohibited by 9.25 are found, it shall be considered as failed. The weld shall show fusion to the root but not necessarily beyond the root, and both legs shall be equal within 1/8 in. (3.2 mm). Convexity shall not exceed the limits specified in 3.6.1, e.g., 1/16 in. (1.6 mm) for a 5/16 in. (8.0 mm) test weld.

5.28.4 Radiographic Test. For acceptable qualification, the weld, as revealed by the radiograph, shall conform to the requirements of 9.25.2.1.

5.28.5 Visual Inspection—Pipe and Tubing. To qualify, the pipe weld, when examined visually, shall conform to the following requirements:

5.28.5.1 The weld shall be free of cracks.

5.28.5.2 All craters shall be filled to the full cross section of the weld.

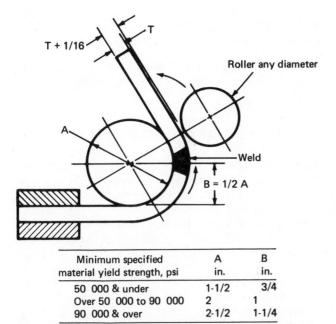

Minimum specified material yield strength, psi	A in.	B in.
50 000 & under	1-1/2	3/4
Over 50 000 to 90 000	2	1
90 000 & over	2-1/2	1-1/4

Fig. 5.27.1B—Alternative wraparound guided-bend test jig

5.28.5.3 The face of the weld shall be at least flush with the outside surface of the pipe, and the weld shall merge smoothly with the base metal. Undercut shall not exceed 1/64 in. (0.4 mm). Weld reinforcement shall not exceed the following:

Pipe wall thickness in. (mm)	Reinforcement, max in.	mm
3/8 (9.5) or less	3/32	2.4
Over 3/8 to 3/4 (19.0) incl.	1/8	3.2
Over 3/4	3/16	4.8

5.28.5.4 The root of the weld shall be inspected, and there shall be no evidence of cracks, incomplete fusion, or inadequate joint penetration. A concave root surface is permitted, within the limits of 5.28.5.5, provided the total weld thickness is equal to or greater than that of the base metal.

5.28.5.5 The maximum root surface concavity shall be 1/16 in. (1.6 mm), and 1/8 in. (3.2 mm) shall be the maximum melt-thru.

5.28.6 Visual Inspection—Plate. For acceptable qualification, the welded test plates, when inspected visually, shall conform to the requirements for visual inspection in 9.25.1.

5.29 Retests

In case a welder fails to meet the requirements of one or more test welds, a retest may be allowed under the following conditions.

5.29.1 An immediate retest may be made consisting of two test welds of each type on which the welder failed. All retest specimens shall meet all the specified requirements.

5.29.2 A retest may be made provided there is evidence that the welder has had further training or practice. In this case, a complete retest (single test welds of each type) shall be made.

5.30 Period of Effectiveness

The welder's qualification as specified in this code shall be considered as remaining in effect indefinitely unless (1) the welder is not engaged in a given process of welding for which he is qualified for a period exceeding six months or unless (2) there is some specific reason to question a welder's ability. In case (1), the requalification test need be made only in the 3/8 in. (9.5 mm) thickness.

5.31 Records

Records of the test results shall be kept by the manufacturer or contractor and shall be available to those authorized to examine them.

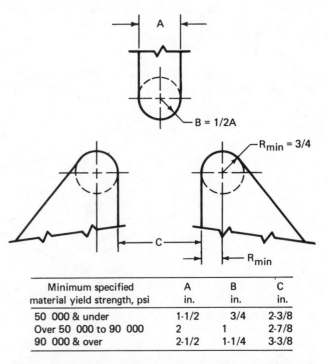

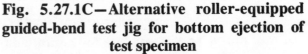

Minimum specified material yield strength, psi	A in.	B in.	C in.
50 000 & under	1-1/2	3/4	2-3/8
Over 50 000 to 90 000	2	1	2-7/8
90 000 & over	2-1/2	1-1/4	3-3/8

Fig. 5.27.1C—Alternative roller-equipped guided-bend test jig for bottom ejection of test specimen

Part D
Welding Operator Qualification

5.32 General

The qualification tests described in Part D are specifically devised tests to determine a welding operator's ability to produce sound welds. The qualification tests are not intended to be used as guides for welding during actual construction. The latter shall be performed in accordance with the requirements of the procedure specification.

5.33 Limitation of Variables

For the qualification of a welding operator, the following rules shall apply.

5.33.1 Qualification established with any one of the steels permitted by this code shall be considered as qualification to weld any of the other steels.

5.33.2 A welding operator qualified with an approved electrode and shielding medium combination shall be considered qualified to weld with any other approved electrode and shielding medium combination for the process used in the qualification test.

5.33.3 For other than electroslag or electrogas welding, a welding operator qualified to weld with multiple electrodes shall be qualified to weld with a single electrode, but not vice versa.

5.33.4 An electroslag or electrogas welding operator qualified with an approved electrode and shielding medium combination shall be considered qualified to weld with any other approved electrode and shielding medium combination for the process used in the qualification test.

5.33.5 A change in the position in which welding of plate is done, as defined in 5.8, shall require requalification.

5.34 Qualification Tests Required

5.34.1 The welding operator qualification test for other than electroslag or electrogas welding shall have a joint detail as follows: 1 in. (25.4 mm) plate, single-V-groove, 20 deg including groove angle, 5/8 in. (15.9 mm) root opening with backing. Backing must be at least 3/8 by 3 in. (9.5 by 76.2 mm) if radiography is used for testing without removal of backing. It must be at least 3/8 in. by 1-1/2 in. (9.5 by 38.1 mm) for mechanical testing or for radiographic testing after the backing is removed. Minimum length of welding groove shall be 15 in. (381 mm) (see Fig. 5.34.1). This test will qualify the welding operator for groove and fillet welding in materials of unlimited thickness for the process and position tested.

5.34.2 The qualification test for an electroslag or electrogas welding operator shall consist of welding a joint of the maximum thickness of material to be used in construction, but the thickness of the material of the test weld need not exceed 1-1/2 in. (38.1 mm) (see Fig. 5.34.2). If a 1-1/2 in. thick test weld is made, no test need be made for lesser thicknesses. This test shall qualify the welding operator for groove and fillet welds in material of unlimited thickness for this process and test position.

5.34.3 The welding operator who makes a complete joint penetration groove weld procedure qualification test that meets the requirements is thereby qualified for that pro-

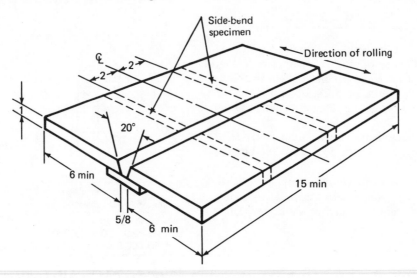

Note: When radiography is used for testing, no tack welds shall be in the test area.

Fig. 5.34.1—Test plate for unlimited thickness—welding operator qualification

cess and test position for plate of thickness equal to or less than the thickness of the test plate welded. If the test plate is 1-1/2 in. (38.1 mm) in thickness in electroslag or electrogas welding or 1 in. (25.4 mm) or over for all other processes, the welding operator will be qualified for all thicknesses. The welding operator is also qualified for fillet welding of plate and pipe for the process and position tested.

5.34.3.1 The welding operator who makes a complete joint penetration groove weld procedure qualification test in pipe that meets the requirements is thereby qualified for that process and test position for pipe. The pipe diameter and wall thickness range qualified for shall be that shown in Table 5.26.1. This qualifies the welding operator for welding groove and fillet welds in plate and pipe as shown in Table 5.23.

5.34.3.2 Qualification of a welding operator on plate in the 1G (flat) or 2G (horizontal) positions shall qualify the welding operator for welding pipe or tubing over 24 in. (600 mm) in diameter in the same welding position.

5.34.4 For fillet weld qualification only: (1) For fillet welds between parts having a dihedral angle (ψ) of 75° or less, the welder shall weld a groove weld test plate as required by 5.34.1; and (2) for joints having a dihedral angle (ψ) greater than 75°, but not exceeding 135°, the welder shall weld a test plate according to Option 1 or Option 2 depending on the contractor's choice, as follows:

5.34.4.1 Option 1. Weld a T-test plate in accordance with Fig. 5.34.4.1.

5.34.4.2 Option 2. Weld a soundness test plate in accordance with Fig. 5.34.4.2.

5.35 Base Metal

The base metal used shall comply with 10.2 or the procedure specification.

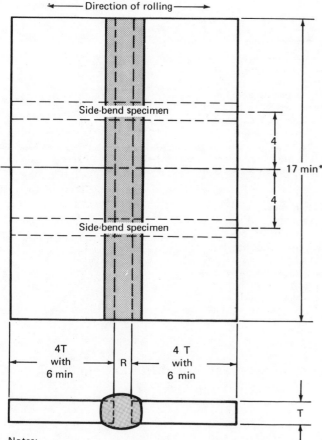

Notes:

1. Root opening "R" established by procedure specification.
2. T = maximum to be welded in construction but need not exceed 1-1/2 in.

   *Extensions need not be used if joint is of sufficient length to provide 17 in. of sound weld.

Fig. 5.34.2—Butt joint for welding operator qualification—electroslag and electrogas welding

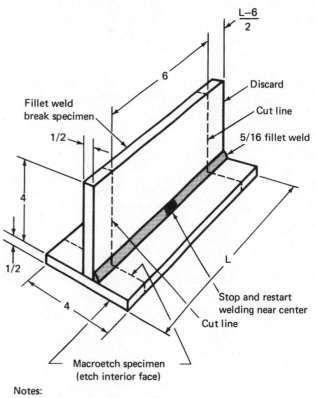

Notes:

1. L = 15 min.
2. Plate thickness and dimensions are minimum.

Fig. 5.34.4.1—Fillet-weld-break and macroetch test plate—welding operator qualification—option 1

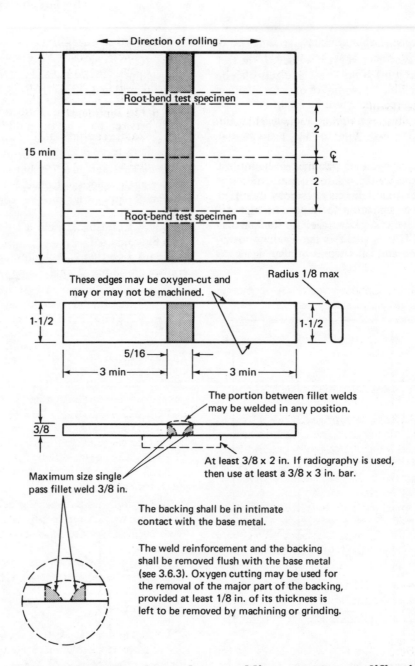

← Direction of rolling →

Root-bend test specimen

15 min

2

₵

2

Root-bend test specimen

These edges may be oxygen-cut and
may or may not be machined.

Radius 1/8 max

1-1/2

1-1/2

5/16

3 min

3 min

The portion between fillet welds
may be welded in any position.

3/8

Maximum size single
pass fillet weld 3/8 in.

At least 3/8 x 2 in. If radiography is used,
then use at least a 3/8 x 3 in. bar.

The backing shall be in intimate
contact with the base metal.

The weld reinforcement and the backing
shall be removed flush with the base metal
(see 3.6.3). Oxygen cutting may be used for
the removal of the major part of the backing,
provided at least 1/8 in. of its thickness is
left to be removed by machining or grinding.

Fig. 5.34.4.2—Fillet-weld root-bend test plate—welding operator qualification—option 2

5.36 Joint Welding Procedure

5.36.1 The welding operator shall follow the joint welding procedure specified by the procedure specification.

5.36.2 Weld cleaning shall be done with the test weld in the same position as the welding position being qualified.

5.37 Test Specimens: Number, Type, and Preparation

5.37.1 For mechanical testing, guided-bend test specimens shall be prepared by cutting the test plate as shown in Figs. 5.34.1, 5.34.2, or 5.34.4.2, whichever is applicable, to form specimens approximately rectangular in cross section. The specimens shall be prepared for testing in accordance with Figs. 5.10.1.3H or 5.10.1.3J, whichever are applicable.

5.37.2 Radiographic Testing

5.37.2.1 At the contractor's option, radiographic testing of the weld may be performed in lieu of the guided-bend test.

5.37.2.2. If radiographic testing is used in lieu of the prescribed bend test, the weld reinforcement need not be ground or otherwise smoothed for inspection unless its surface irregularities or juncture with the base metal would cause objectionable weld discontinuities to be obscured in the radiograph. If the backing is removed for radiographic testing, the root shall be ground flush with the base metal (see 3.6.3).

5.37.3 The fillet weld break and macroetch test specimens shall be cut from the test joint as shown in Fig. 5.34.4.1. The end of the macroetch test specimen shall be smooth for etching.

5.38 Method of Testing Specimens

5.38.1 Root- or Side-Bend Specimens. Each specimen shall be bent in a jig having the contour shown in Fig. 5.27.1 and otherwise shall be substantially in accordance with that figure. Any convenient means may be used to move the plunger member with relation to the die member. The specimen shall be placed on the die member of the jig with the weld at midspan. Side-bend specimens shall be placed with that side showing greater discontinuities, if any, directed toward the gap; root-bend (fillet weld soundness) specimens shall be placed with the root of the weld directed toward the gap.

5.38.2 The radiographic procedure and technique shall be in accordance with the requirements of Part B, Section 6. Only the center half of the length of the test plate shall be subject to testing.

5.38.3 Fillet-Weld-Break Test. The entire length of the fillet weld shall be examined visually and then a 6 in. (152 mm) long specimen shall be loaded in such a way that the root of the weld is in tension. The load shall be steadily increased or repeated until the specimen fractures or bends flat upon itself.

5.38.4 Macroetch Test. The test specimens shall be prepared with a finish suitable for macroetch examination. A suitable solution shall be used to give a clear definition of the weld.

5.39 Test Results Required

5.39.1 Root- or Side-Bend Tests. The convex surface of the specimen shall be examined for the appearance of cracks or other open discontinuities. Any specimen in which a crack or other open discontinuity exceeding 1/8 in. (3.2 mm) measured in any direction is present after bending shall be considered as having failed. Cracks occurring on the corners of the specimen during testing shall not be considered.

5.39.2 Radiographic Test. For acceptable qualification, the weld as revealed by the radiograph shall conform to the requirements of 9.25.2.1.

5.39.3 Fillet-Weld-Break Test

5.39.3.1 To pass the visual examination, the fillet weld shall present a reasonably uniform appearance and shall be free of overlap, cracks, and excessive undercut. There shall be no porosity visible on the surface of the weld.

5.39.3.2 The specimen shall pass the test if it bends flat upon itself. If the fillet weld fractures, the fractured surface shall show complete fusion into the root of the joint and shall exhibit no inclusion or porosity larger than 3/32 in. (2.4 mm) in the greatest dimension. The sum of the greatest dimensions of all inclusions and porosity shall not exceed 3/8 in. (9.5 mm) in the 6 in. (152 mm) long specimen.

5.39.4 Macroetch Test. The test specimen shall be examined for discontinuities and if discontinuities prohibited by 9.25 are found, it shall be considered as failed. The weld shall show fusion to the root but not necessarily beyond the root and both legs shall be equal within 1/8 in. (3.2 mm). Convexity shall not exceed the limits specified in 3.6.1, 1/16 in. (1.6 mm) for a 5/16 in. (8.0 mm) test weld, for example.

5.39.5 Visual Inspection. For acceptable qualification:

(1) The welded test plate, when inspected visually, shall conform to the requirements for visual inspection in 9.25.1;

(2) The pipe weld, when inspected visually, shall conform to the requirements of 5.28.5.

5.40 Retests

If a welding operator fails to meet the requirements of one or more test welds, a retest may be allowed under the following conditions.

5.40.1 An immediate retest may be made consisting of two test welds of each type on which he failed. All specimens shall meet all the requirements specified for such welds.

5.40.2 A retest may be made provided there is evidence that the welding operator has had further training or practice. In this case a complete retest (single test welds of each type) shall be made.

5.41 Period of Effectiveness

The welding operator's qualification specified in Part D shall be considered as remaining in effect indefinitely unless (1) the welding operator is not engaged in the given process of welding for which he is qualified for a period exceeding six months; or unless (2) there is some specific reason to question the welding operator's ability.

5.42 Records

Records of the test results shall be kept by the manufacturer or contractor and shall be available to those authorized to examine them.

Part E
Qualification of Tackers

5.43 General

The qualification tests described in Part E are specially devised tests to determine a tacker's ability to produce sound welds. The qualification tests are not intended to be used as a guide for tack welding during actual construction. The latter shall be performed in accordance with the requirements of the procedure specification.

5.44 Limitation of Variables

For the qualification of a tacker the following rules shall apply.

5.44.1 Qualification established with any one of the steels permitted by this code shall be considered as qualification to tack weld any of the other steels.

5.44.2 A tacker qualified for shielded metal arc welding with an electrode identified in Table 5.44.2 shall be considered qualified to tack-weld with any other electrode in the same group designation and with any electrode listed in a numerically lower group designation.

Table 5.44.2
Electrode classification groups—
tacker qualification

Group designation	AWS electrode classification*
F4	EXX15, EXX16, EXX18
F3	EXX10, EXX11
F2	EXX12, EXX13, EXX14
F1	EXX20, EXX24, EXX27, EXX28

*The letters "XX" used in the classification-designation in this table stand for the various strength levels (60, 70, 80, 90, 100, 110, and 120) of electrodes.

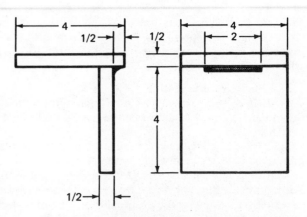

Fig. 5.47—Fillet-weld-break specimen—tacker qualification

5.44.3 A tacker qualified with an approved electrode and shielding medium combination shall be considered qualified to tack weld with any other approved electrode and shielding medium combination for the process used in the qualification test.

5.44.4 A tacker shall be qualified for each process used.

5.44.5 A change in the position in which tacking is done, as defined in 5.8, shall require requalification.

5.45 Qualification Tests Required

A tacker shall be qualified by one test plate made in each position in which he is to tack weld.

5.46 Base Metal

The base metal used shall comply with 10.2 or the procedure specification.

5.47 Test Specimens: Number, Type, and Preparation

The tacker shall make a 1/4 in. (6.4 mm) maximum size tack weld approximately 2 in. (50.8 mm) long on the fillet-weld-break specimen as shown in Fig. 5.47, using a 5/32 in. (4.0 mm) diameter electrode.

5.48 Method of Testing Specimens

A force shall be applied to the specimen as shown in Fig. 5.48 until rupture occurs. The force may be applied by any convenient means. The surface of the weld and of the fracture shall be examined visually for defects.

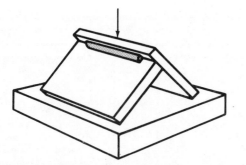

Fig. 5.48—Method of rupturing specimen—tacker qualification

5.49 Test Results Required

5.49.1 The tack weld shall present a reasonably uniform appearance and shall be free of overlap, cracks, and excessive undercut. There shall be no porosity visible on the surface of the tack weld.

5.49.2 The fractured surface of the tack weld shall show fusion to the root but not necessarily beyond and shall exhibit no incomplete fusion to the base metal nor any inclusion or porosity larger than 3/32 in. (2.4 mm) in greatest dimension.

5.49.3 A tacker who passes the fillet weld break test shall be eligible to tack-weld all types of joints for the process and in the positions in which he has qualified.

5.50 Retests

In case of failure to pass the above test, the tacker may make one retest without additional training.

5.51 Period of Effectiveness

A tacker who passes the test just described shall be considered eligible to perform tack welding indefinitely in the positions and with the process for which he is qualified unless there is some specific reason to question his ability. In such case, the tacker shall be required to demonstrate his ability to make sound tack welds by again passing the prescribed tack welding test.

5.52 Records

Records of the test results shall be kept by the manufacturer or contractor and shall be available to those authorized to examine them.

6. Inspection

Part A
General Requirements

6.1 General

6.1.1 The Inspector[29] designated by the Engineer shall ascertain that all fabrication by welding is performed in accordance with the requirements of this code.

6.1.2 He shall be furnished with complete detail drawings showing the size, length, type, and location of all welds to be made.

6.1.3 He shall be notified, in advance, of the start of any welding operations.

6.2 Inspection of Materials

The Inspector shall make certain that only materials conforming to the requirements of this code are used.

6.3 Inspection of Welding Procedure Qualification and Equipment

6.3.1 The Inspector shall make certain that all welding procedures are prequalified and covered by a welding procedure specification or are qualified in accordance with 5.2 of this code.

6.3.2 He shall inspect the welding equipment to be used for the work to make certain that it conforms to the requirements of 3.1.2.

---

29. The Inspector is the duly designated person who acts for and in behalf of the Engineer in all inspection and quality matters within the scope of this code.

6.4 Inspection of Welder, Welding Operator, and Tacker Qualifications

6.4.1 The Inspector shall permit welding to be performed only by welders, welding operators, and tackers who are qualified in accordance with the requirements of 5.3, or shall make certain that each welder, welding operator, or tacker has previously demonstrated his qualification under other acceptable supervision.

6.4.2 When the quality of a welder's, welding operator's, or tacker's work appears to be below the requirements of this code, the Inspector may require that welder, welding operator, or tacker to demonstrate his ability to produce sound welds by means of a simple test, such as the fillet weld break test, or by requiring complete requalification in accordance with 5.3.

6.4.3 The Inspector shall require requalification of any welder, welding operator, or tacker who has not used the process for which he has been qualified for a period exceeding six months.

6.5 Inspection of Work and Records

6.5.1 The Inspector shall make certain that the size, length, and location of all welds conform to the requirements of this code and to the detail drawings and that no unspecified welds have been added without approval.

6.5.2 The Inspector shall make certain that only welding procedures which meet the provisions of 5.1 and 5.2 are employed.

6.5.3 The Inspector shall make certain that electrodes are used only in the positions and with the type of welding current and polarity for which they are classified.

6.5.4 The Inspector shall, at suitable intervals, observe the technique and performance of each welder, welding operator, and tacker to make certain that the applicable requirements of Section 4 are met.

6.5.5 The Inspector shall examine the work to make certain that it meets the requirements of Section 3 and 8.15, 9.25, or 10.17, as applicable. Size and contour of welds shall be measured with suitable gages. Visual inspection for cracks in welds and base metal and other discontinuities should be aided by a strong light, magnifiers, or such other devices as may be found helpful.

6.5.6 The Inspector shall identify with a distinguishing mark all parts or joints that he has inspected and accepted.

6.5.7 The Inspector shall keep a record of qualifications of all welders, welding operators, and tackers; all procedure qualifications or other tests that are made; and such other information as may be required.

6.6 Obligations of Contractor

6.6.1 The contractor shall comply with all requests of the Inspector to correct improper workmanship and to remove and replace or correct, as instructed, all welds found unacceptable or deficient.

6.6.2 In the event that faulty welding or removal for rewelding damages the base metal so that, in the judgment of the Engineer, its retention is not in accordance with the intent of the drawings and specifications, the contractor shall remove and replace the damaged base metal or shall compensate for the deficiency in a manner approved by the Engineer.

6.6.3 The contractor shall be responsible for visual examination and necessary correction of all welds in accordance with the requirements of 3.7 and 8.15.1, 9.25.1, or 10.17.1.

6.6.4 When nondestructive testing other than visual inspection is specified in the information furnished to bidders, it shall be the contractor's responsibility to insure that all specified welds meet the quality requirements of 8.15, 9.25, or 10.17, whichever is applicable.

6.6.5 If nondestructive testing other than visual inspection is not specified in the original contract agreement but is subsequently requested by the owner, the contractor shall perform any requested testing or shall permit any testing to be performed in accordance with 6.7. The owner shall be responsible for all associated costs including handling, surface preparation, nondestructive testing, and repair of discontinuities other than those listed in 8.15.1, 9.25.1, or 10.17.1, whichever is applicable, at rates mutually agreeable between owner and contractor. However, if such testing should disclose an attempt to defraud or gross nonconformance to this code, repair work shall be done at the contractor's expense.

6.7 Nondestructive Testing

6.7.1 When nondestructive testing other than visual is to be required, it shall be so stated in information furnished to the bidders. This information shall designate the welds to be examined, the extent of examination of each weld, and the method of testing.

6.7.2 Welds tested nondestructively that do not meet the requirements of this code shall be repaired by the methods permitted by 3.7.

6.7.3 When radiographic testing is used, the procedure and technique shall be in accordance with Part B of this section.

6.7.4 When ultrasonic testing is used, the procedure and technique shall be in accordance with Part C of this section.

6.7.5 When magnetic particle testing is used, the procedure and technique shall be in accordance with ASTM E109, and the standards of acceptance shall be in accordance with 8.15, 9.25, or 10.1.7 of this code, whichever is applicable.

6.7.6 For detecting discontinuities that are open to the surface, dye penetrant inspection may be used. The standard methods set forth in ASTM E165 shall be used for dye penetrant inspection, and the standards of acceptance shall be in accordance with 8.15, 9.25, or 10.17 of this code, whichever is applicable.

6.7.7 Personnel Qualification. Personnel performing nondestructive testing shall be qualified in accordance with the current edition of American Society for Nondestructive Testing Recommended Practice No. SNT-TC-1A.[30] Only individuals qualified for NDT LEVEL I and working under the NDT LEVEL II or individuals qualified for NDT LEVEL II may perform nondestructive testing.

Part B
Radiographic Testing of Welds

6.8 General

6.8.1 The procedure and standards set forth in Part B are to govern radiographic testing of welds when such inspection is required by the stipulation of 6.7. These procedures are entirely for testing groove welds in butt joints.

---

30. Available from the American Society for Nondestructive Testing, 3200 Riverside Drive, Columbus, OH 43221.

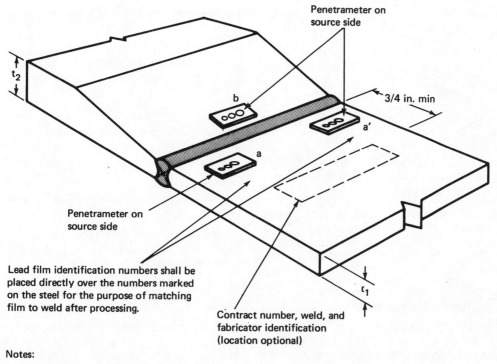

Notes:

1. a, a' = 2% t_1.
2. b = 2% t_2 but not more than required by 2t_1.

Fig. 6.10.4—Radiograph identification and penetrameter location on transition joints

6.8.2 Variation in testing procedure, equipment, and acceptance standards not included in Part B may be used upon agreement with the Engineer. Such variations include the radiographic testing of fillet, T-, or corner welds; changes in source-to-film distances; unusual application of film for unusual geometries; unusual penetrameter application; film types or densities; and film exposure or development variations.

6.9 Extent of Testing

6.9.1 Information furnished to the bidder shall clearly identify the extent of radiographic testing.

6.9.2 When complete testing is specified, the entire length of the weld in each designated joint shall be inspected.

6.9.3 When spot testing is specified, the number of spots in each designated category of welded joint to be radiographed in a stated length of weld shall be included in information furnished to the bidders. Each spot radiograph shall show at least 4 in. (102 mm) of weld length. If a spot radiograph shows discontinuities that require repair, as defined in 6.11.1, two adjacent spots shall be inspected. If discontinuities requiring repair are shown in either of these, the entire length of weld in that welded joint shall be tested radiographically.

6.10 Radiographic Procedure

6.10.1 Radiographs shall be made by either x-ray or isotope radiation methods. All radiographs shall determine quantitatively the size of discontinuities having thickness equal to or greater than 2% of the thickness of the thinner of the parts joined by the weld under examination. They shall be clean, free of film processing defects, and shall have an H&D[31] density of not less than 1.5 nor more than 4.0. Although radiographs (each single film) may have an H&D density of 1.5 minimum to 4.0 maximum, densities within the range of 2.5 to 3.5 are preferred. Radiographs, except as modified by 6.10.4, shall show

6.10.1.1 The smallest hole in each penetrameter as specified by Fig. 6.10.5B.

6.10.1.2 The penetrameter identification number.

6.10.1.3 The radiographic identification and location numbers indicated in Figs. 6.10.4 and 6.10.5A and required by 6.10.6.

6.10.2 Radiography shall be performed in accordance with all applicable safety requirements.

---

31. The film characteristic curve (sensitometric curve) that expresses the relation between the exposure applied to photographic material and the resulting photographic density.

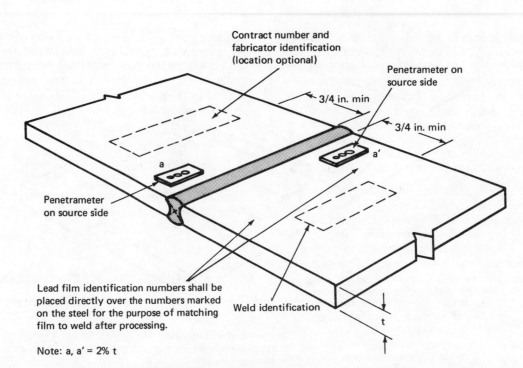

Contract number and
fabricator identification
(location optional)

Penetrameter on
source side

3/4 in. min

3/4 in. min

Penetrameter on source side

a

a'

t

Lead film identification numbers shall be
placed directly over the numbers marked
on the steel for the purpose of matching
film to weld after processing.

Weld identification

Note: a, a' = 2% t

Fig. 6.10.5A—Radiograph identification and penetrameter location on approximately equal thickness joints

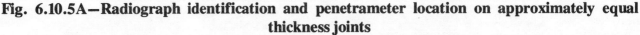

T = Thickness of the penetrameter

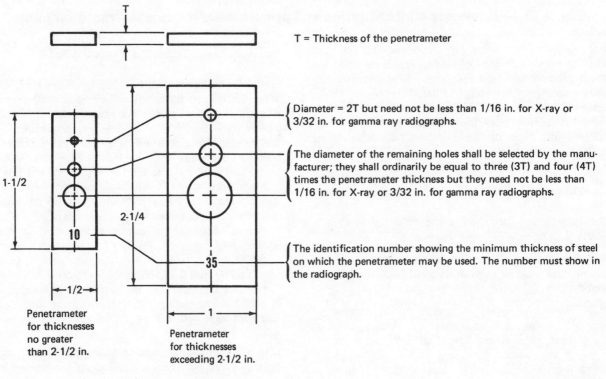

Diameter = 2T but need not be less than 1/16 in. for X-ray or 3/32 in. for gamma ray radiographs.

The diameter of the remaining holes shall be selected by the manufacturer; they shall ordinarily be equal to three (3T) and four (4T) times the penetrameter thickness but they need not be less than 1/16 in. for X-ray or 3/32 in. for gamma ray radiographs.

The identification number showing the minimum thickness of steel on which the penetrameter may be used. The number must show in the radiograph.

1-1/2

1/2

10

2-1/4

35

1

Penetrameter
for thicknesses
no greater
than 2-1/2 in.

Penetrameter
for thicknesses
exceeding 2-1/2 in.

Notes:

1. All holes shall be true and normal to the surface and not chamfered.

2. Penetrameters shall be made of carbon steel or Type 304 stainless steel.

Fig. 6.10.5B—Details of penetrameters

6.10.3 A weld that is to be radiographed need not be ground or otherwise smoothed for purposes of radiographic testing unless its surface irregularities or juncture with the base metal could cause objectionable weld discontinuities to be obscured in the radiograph. When weld reinforcement or backing is not removed, carbon steel shims shall be placed under the penetrameter so that the total thickness of steel between the penetrameter and the film is at least equal to the average thickness of the weld measured through its reinforcement and backing.

6.10.4 When weld transitions in thickness are radiographed and the ratio of the thicker weld section to the thinner weld section is 3 or greater, radiographs should be exposed to produce a density of 3.0 to 4.0 in the thinner section. When this is done, densities of less than 1.5 will be accepted in the thicker section. Except for this condition, densities outside the maximum and minimum limits specified in 6.10.1 shall be cause for rejection of the film. Penetrameters on transition joints shall be positioned as shown in Fig. 6.10.4.

6.10.5 Two or more penetrameters shall be used for each radiograph on a film 10 in. (254 mm) or more in length. Only one penetrameter need be used for radiographs on films less than 10 in. in length. Penetrameters shall be placed on the side of the work nearer the radiation source, as shown in Fig. 6.10.4 and Fig. 6.10.5A. Penetrameters shall conform to the details shown in Fig. 6.10.5B except that other penetrameters, such as ASME, may be used provided they have identification numbers indicating penetrameter thickness in thousandths of an inch and comply with all other conditions of this paragraph and Fig. 6.10.5B. The thickness of each penetrameter shall be equal to or less than 2% of the thickness of the thinner of the parts joined by the weld under examination, but need not be less than 0.005 in. (0.13 mm).

6.10.5.1 Each penetrameter shall carry lead numbers which identify the minimum thickness of material (to the nearest 0.05 in. [1.3 mm]) for which it may be used. The images of these identifying numbers shall appear clearly on the radiograph.

6.10.6 A radiograph identification mark and two location identification marks shall be placed on the steel at each radiograph location. A corresponding radiograph identification number and a location identification number, both of which will show in the radiograph, shall be superimposed on each of the location identification marks made on the steel to provide a means for matching the developed radiograph to the weld. Any additional information shall be preprinted no less than 3/4 in. (19.0 mm) from the edge of the weld or shall be indicated by lead figures on steel.

6.10.7 Radiographs shall be made with a single source of radiation approximately centered with respect to the length of area being examined. The perpendicular distance from the radiation source to the film shall be no less than seven times the maximum thickness of the weld under examination, and the rays shall not penetrate the weld at an angle greater than 26-1/2 deg from a line perpendicular to the weld surface. The film, during exposure, shall be as close to the surface of the weld opposite the source of radiation as possible.

6.11 Acceptability of Welds

Welds shown by radiographic testing to have discontinuities prohibited by 8.15.2, 9.25.2, or 10.17.2 shall be corrected in accordance with 3.7.

6.12 Examination, Report, and Disposition of Radiographs

6.12.1 The contractor shall provide a suitable high intensity viewer with sufficient capacity to illuminate radiographs with a density of 4.0 without difficulty. (It is recommended that at least a two-level or variable-intensity illuminator be used.)

6.12.2 Before a weld subject to radiographic testing by the contractor for the owner is accepted, all of its radiographs, including any that show unacceptable quality prior to repair, and a report interpreting them, shall be submitted to the Inspector.

6.12.3 A full set of radiographs for welds subject to radiographic testing by the contractor for the owner, including any that show unacceptable quality prior to repair, shall be delivered to the owner upon completion of the work. The contractor's obligation to retain radiographs shall cease (1) upon delivery of this full set to the owner or (2) one full year after completion of the contractor's work, in the event that delivery is not made.

Part C
Ultrasonic Testing of Groove Welds

6.13 General

6.13.1 The procedures and standards set forth in Part C are to govern the ultrasonic testing of groove welds between the thicknesses of 5/16 in. (8.0 mm) and 8 in. (203 mm) inclusive, when such testing is required by 6.7 of this code. These procedures and standards are not to be used for testing tube-to-tube T-, Y-, or K-connections (see 10.17.4), or as a basis for rejection of the base metal.

6.13.2 Variations in testing procedure, equipment, and acceptance standards not included in Part C of Section 6 may be used upon agreement with the Engineer. Such variations include other thicknesses, weld geometries, transducer sizes, frequencies, couplant, etc.

6.13.3 To detect possible piping porosity, spot radiography is suggested to supplement ultrasonic testing of electroslag and electrogas butt welds in material 2 in. (50.8 mm) and over in thickness.

6.14 Extent of Testing

6.14.1 Information furnished to the bidders shall clearly identify the extent of ultrasonic testing.

6.14.2 When complete testing is specified, the entire length of the weld in each designated joint shall be tested.

6.14.3 When spot testing is specified, the number of spots in each designated category of weld or the number required to be made in a stated length of weld shall be included in the information furnished to the bidders. Each spot tested shall cover at least 4 in. (102 mm) of the weld length. When spot testing reveals discontinuities that require repair, two adjacent spots shall be tested. If discontinuities requiring repair are revealed in either of these, the entire length of the weld in that welded joint shall be tested ultrasonically.

6.15 Ultrasonic Equipment

6.15.1 The ultrasonic test instrument shall be of the pulse-echo type. It shall generate, receive, and present on a cathode ray tube (CRT) screen pulses in the frequency range from one to six megahertz (MHz). The presentation on the CRT screen shall be the "video" type, characterized by a clean, crisp trace.

6.15.2 The horizontal linearity of the test instrument shall be within plus or minus 5% over the linear range which includes 90% of the sweep length presented on the CRT screen for the longest sound path to be used. The horizontal linearity shall be measured by the techniques prescribed by Section 7.9 of ASTM E317 except that the results may be tabulated rather than graphically presented.

6.15.3 Test instruments shall include internal stabilization so that after warm up, no variation in response greater than ±1 dB occurs with supply voltage change of +15% nominal or, in the case of battery powered instruments, over the battery charge operating life. There shall be an alarm or meter to signal a drop in battery voltage prior to instrument shutoff due to battery exhaustion.

6.15.4 The test instrument shall have a calibrated gain control (attenuator) adjustable in discrete 1 or 2 dB steps over a range of at least 60 dB. The accuracy of the gain control settings shall be within plus or minus 1 dB.

6.15.5 The dynamic range of the instrument's CRT display shall be such that a difference of 1 dB of amplitude can be easily detected on the CRT.

6.15.6 Straight beam search unit transducers shall have an active area of not less than 1/2 in.² (323 mm²) nor more than 1 in.² (645 mm²). The transducer shall be round or square. Transducer frequency shall be 2 to 2.5 MHz. Transducers shall be capable of resolving the three reflections as described in 6.21.1.3.

6.15.7 Angle beam search units shall consist of a transducer and an angle wedge. The unit may be comprised of the two separate elements or may be an integral unit.

6.15.7.1 The transducer frequency shall be between 2 and 2.5 MHz, inclusive.

6.15.7.2 The transducer crystal may vary in size from 1/2 to 1 in. (12.7 to 25.4 mm) in width and from 1/2 to 13/16 in. (12.7 to 20.6 mm) in height (see Fig. 6.15.7.2).

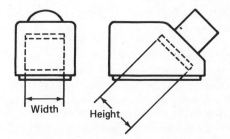

Fig. 6.15.7.2—Transducer crystal

6.15.7.3 The search unit shall produce a sound beam in the material being tested within plus or minus 2 deg of the following proper angles: 70 deg, 60 deg, or 45 deg, as described in 6.22.2.2.

6.15.7.4 Each search unit shall be marked to clearly indicate the frequency of the transducer, nominal angle of refraction, and index point. The index point location procedure is described in 6.21.2.1.

6.15.7.5 Internal reflections from the search unit, with a screen presentation higher than the horizontal reference line, appearing on the screen to the right of the sound entry point shall not occur beyond 1/2 in. (12.7 mm) equivalent distance in steel when the sensitivity is set as follows: 20 dB more than that required to produce a maximized horizontal reference line height indication from the 0.06 in. (1.52 mm) diameter hole in the International Institute of Welding (IIW) reference block (see Fig. 6.16.1A).

6.15.7.6 The dimensions of the search unit shall be such that the closeness of approach to the weld reinforcement shall not exceed the requirements of 6.21.2.6. The search unit shall be positioned for maximum indication from the 0.06 in. (1.52 mm) diameter hole in the IIW calibration block.

6.15.7.7 The combination of search unit and instrument shall resolve three holes in the resolution test block shown in Fig. 6.15.7.7. The search unit position is described in 6.21.2.5. The resolution shall be evaluated with the instrument controls set at normal test settings and with indications from the holes brought to mid-screen height. Resolution shall be sufficient to distinguish at least the peaks of indications from the three holes.

6.16 Calibration Standards

6.16.1 The International Institute of Welding (IIW) ultrasonic reference block, shown in Fig. 6.16.1A, shall be the standard used for both distance and sensitivity calibration. More portable reference blocks of other design may be used provided they meet the requirements of this specification and are referenced back to the IIW block. Approved designs are shown in Fig. 6.16.1B. See Fig. 6.21 for applications.

6.16.2 The use of a "corner" reflector for calibration purposes is prohibited.

6.17 Equipment Calibration

6.17.1 The instrument's gain control (attenuator) shall meet the requirements of 6.15.5 and shall be checked for correct calibration at two-month intervals in accordance with a procedure approved by the manufacturer of the instrument.

6.17.2 Horizontal linearity shall be checked by the techniques prescribed in 6.15.2 after each 40 hours of instrument use.

6.17.3 With the use of an approved calibration block, each angle beam search unit shall be checked after each eight hours of use to determine that the contact face is flat, that the sound entry point is correct, and that the beam angle is within the permitted plus or minus 2 deg tolerance. Search units which do not meet these requirements shall be corrected or replaced.

6.18 Calibration for Testing

6.18.1 Calibration for sensitivity and horizontal sweep (distance) shall be made by the ultrasonic operator just prior to and at the location of testing of each weld and at intervals of 30 min as testing proceeds. Recalibration shall be made each time there is a change of operators, when transducers are changed, when new batteries are installed, or when equipment operating from a 110 volt source is connected to a different power outlet.

6.18.2 Calibration for straight beam testing shall be performed as follows:

6.18.2.1 The horizontal sweep shall be adjusted for distance calibration to present the equivalent of at least two plate thicknesses on the CRT screen.

6.18.2.2 The sensitivity shall be adjusted at a location free of indications so that the first back reflection from the far side of the plate will be 50 to 75% of full screen height (6.21.1.2). For this purpose, the reject (clipping) control shall be turned off.

6.18.3 Calibration for angle beam testing shall be performed as follows:

6.18.3.1 The horizontal sweep shall be adjusted to represent the actual sound path distance by using acceptable distance calibration blocks shown in Figs. 6.16.1A and 6.16.1B. This distance calibration shall be made using either the 5 in. (127 mm) scale or 10 in. (254 mm) scale on the CRT screen, whichever is appropriate, unless joint configuration or thickness prevents full examination of the weld at either of these settings. The search unit position is described in 6.21.2.3.

6.18.3.2 With the unit adjusted to conform to the requirements of 6.15, the sensitivity shall be adjusted by the use of the gain control (attenuator) so that a horizontal reference level trace deflection results on the CRT screen with the maximum indication from the 0.06 in. (1.52 mm) diameter hole in the IIW block or from the equivalent reference reflector in other acceptable calibration blocks. The search unit position is described in 6.21.2.4. This basic sensitivity then becomes the zero reference level for discontinuity evaluation and shall be recorded on the ultrasonic test reports under reference level. See Appendix E for a sample ultrasonic test report form.

6.19 Testing Procedure

6.19.1 A "Y" accompanied with a weld identification number shall be clearly marked on the base metal adjacent to the weld at the left end of each weld that is ultrasonically tested. This identification number serves as an orientation direction for weld discontinuity location and as the report number on the report form. (See Appendix E for suggested report form.)

6.19.2 All surfaces to which a search unit is applied shall be free of weld spatter, dirt, grease, oil (other than that used as a couplant), and loose scale and shall have a contour permitting intimate coupling. Tight layers of paint need not be removed unless the thickness exceeds 10 mils (0.25 mm).

6.19.3 A couplant shall be used between the search unit and the metal. The couplant shall be either glycerin with a wetting agent added, if needed, or a cellulose gum and water mixture of a suitable consistency. Light machine oil or equivalent may be used for couplant on calibration blocks.

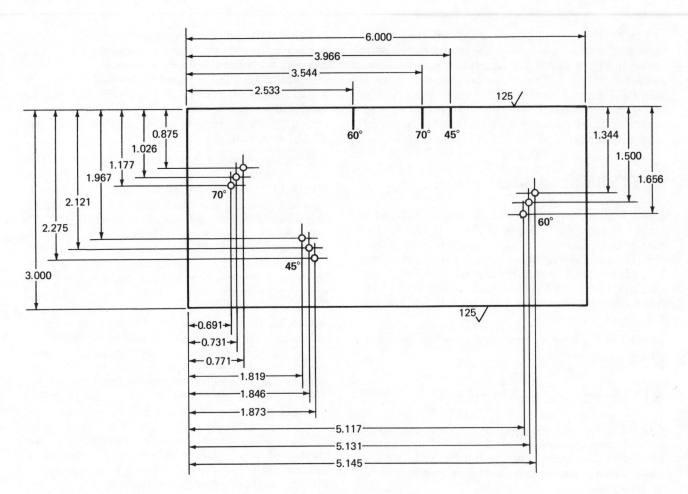

Notes:

1. The test block is 3 in. x 1 in. x 6 in. Finish all over to a maximum of 125 μin. rms.

2. Material - ASTM A36 or equivalent.

3. 1/16 in. diameter holes are to be drilled at 90 degrees to the surface.

4. Degree lines are to be scribed on the surface as shown.

5. Degree numbers are to be stenciled on the surface as shown.

Fig. 6.15.7.7—Resolution test block

6.19.4 The entire base metal through which ultrasound must travel to test the weld shall be tested for laminar reflectors using a straight beam search unit conforming to the requirements of 6.15.6 and calibrated in accordance with 6.18.2. If any area of base metal exhibits total loss of back reflection and is located in a position that would interfere with the normal weld scanning procedure, the following alternative weld scanning procedure shall be used.

6.19.4.1 The area of the laminar reflector and its depth from the surface shall be determined and reported on the ultrasonic test report.

6.19.4.2 If part of a weld is inaccessible to testing in accordance with the requirements of Table 6.19.5.2, due to laminar content recorded in accordance with 6.19.4.1, the testing shall be conducted (1) using an alternative scanning pattern shown in Fig. 6.22, or (2) by first grinding the weld surfaces flush to make total weld areas accessible to ultrasonic testing, or both.

6.19.5 Welds shall be tested using an angle beam search unit conforming to the requirements of 6.15.7 with the instrument calibrated in accordance with 6.18.3 using the angle as shown in Table 6.19.5.2. Following calibration and during testing, the only instrument adjustment permitted is in the sensitivity level adjustment with the calibrated gain control or attenuator. Sensitivity shall be

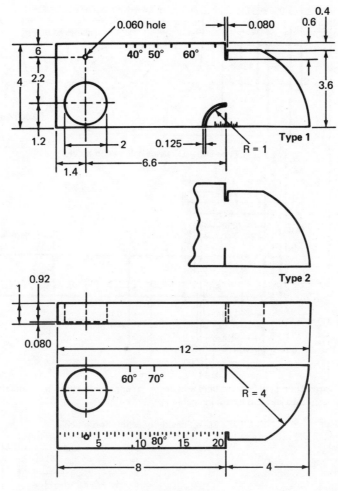

Notes:

1. Other IIW-approved reference blocks with slightly different dimensions or distance calibration slot features are permissible.

2. Material: ASTM A36 steel or equivalent.

Fig. 6.16.1A—International Institute of Welding (IIW) ultrasonic reference block

increased from the reference level for weld scanning in accordance with Table 8.15.3 or 9.25.3, as applicable.

6.19.5.1 If mechanically possible, all welds shall be scanned from both sides on the same face for longitudinal and transverse discontinuities. The applicable scanning pattern or patterns shown in Fig. 6.22 shall be used.

6.19.5.2 The testing angle shall be as shown in Table 6.19.5.2, and the transducer size must conform to 6.15.7.2.

6.19.5.3 When a discontinuity indication appears on the screen, the maximum attainable indication from the discontinuity shall be adjusted to produce a horizontal reference level trace deflection on the CRT screen. This adjustment shall be made with the calibrated gain control or attenuator, and the instrument reading, in decibels, shall be recorded on the ultrasonic test report under the heading "Indication Level."

6.19.5.4 The "Attenuation Factor," "c," on the test report is attained by subtracting 1 in. (25.4 mm) from the sound path distance and multiplying the remainder by two.

6.19.5.5 The "Indication Rating," "d," on the test report is the difference between the "Reference Level" and the "Indication Level" after the "Indication Level" has been corrected by the "Attenuation Factor."

Instruments with gain in dB: $a - b - c = d$
Instruments with attenuation in dB: $b - a - c = d$

6.19.5.6 The length of a discontinuity as entered under "Indication Length" on the test report shall be determined by locating points at each end at which the indication amplitude drops 6 dB and measuring between the points

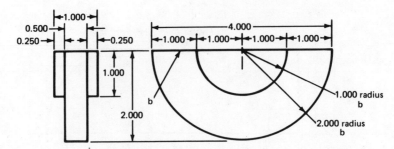

Type DC — Distance calibration block

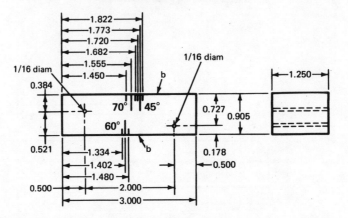

Note: Sound entry point lines and degree of angle indications to be indented into surfaces where indicated.

Type SC — Sensitivity calibration block

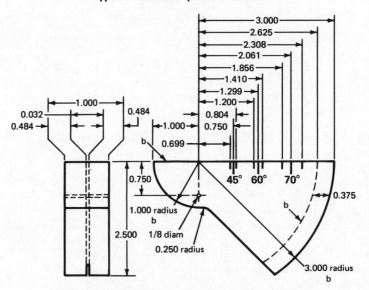

Type DSC — Distance and sensitivity calibration block

Notes:

1. Material: ASTM A36 or equivalent.

2. Dimension tolerance — ± 0.005.

3. Minimum surface finish — b — 64 to 125 μin. rms.
    others — 125 to 250 μin. rms.

Fig. 6.16.1B—Other calibration blocks

Table 6.19.5.2
Testing angle

Procedure chart

Material thickness, in.

Weld type	5/16 to 1-1/2		>1-1/2 to 1-3/4		>1-3/4 to 2-1/2		>2-1/2 to 3-1/2		>3-1/2 to 4-1/2		>4-1/2 to 5		>5 to 6-1/2		>6-1/2 to 7		>7 to 8	
	*		*		*		*		*		*		*		*		*	
Butt	1	0	1	F	1G or 4	F	1G or 5	F	6 or 7	F	8 or 10	F	9 or 11	F	12 or 13	F	12	F
T	1	0	1	F or XF	4	F or XF	5	F or XF	7	F or XF	10	F or XF	11	F or XF	12 or 13	F or XF	—	—
Corner	1	0	1	F or XF	1G or 4	F or XF	1G or 5	F or XF	6 or 7	F or XF	8 or 10	F or XF	9 or 11	F or XF	F or 14	F or XF	—	—
Electrogas & electroslag	1	0	1	0	1G or 4	1**	1G or 5	P1 or P3	6 or 7	P3	11 or 15	P3	11 or 15	P3	11 or 15	P3	11 or 15**	P3

Note: All examinations are to be made from Face "A" unless noted and scanned from both sides of weld (on Face "A") where mechanically possible. All examinations are to be made in Leg I where possible or Leg II only when necessary to test weld areas made inaccessible by unground weld surface contour. A maximum of Leg III is to be used only where extra weld bead width prevents scanning of certain weld areas in Leg I or Leg II. (See Glossary of Terms, Appendix I.)

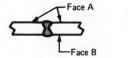
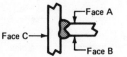
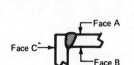

Pitch and catch

Note: Procedure G, 6, 8, 9, 12, 14, or 15 must be followed when testing welds which have been ground flush. The need for grinding may either be to satisfy contract requirements or at the option of the contractor to provide a more favorable working condition. Face "A" on both connecting members must lie in a single plane.

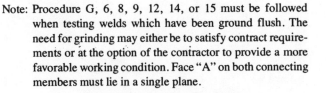

Example: Butt weld in 4 in. base metal
No. 6 procedure

See Legend and Notes on page 114.

Table 6.19.5.2 (continued)
Testing angle

Legend:

   X—Check from Face "C."

   G—Grind weld face flush.

   O—Not required.

A Face—the face of the material from which the initial scanning is done (on T- and corner joints, follow above sketches).

B Face—opposite the "A" face (same plate).

C Face—the face opposite the weld on the connecting member of a T- or corner joint.

   *Required only where reference level indication of discontinuity is noted in fusion zone while searching at scanning level with primary procedure selected from first column.

   **Use 15 or 20 in. screen distance calibration.

P—Pitch and catch shall be conducted for further discontinuity evaluation in only the middle half of the material thickness with only 45 deg or 70 deg transducers of equal specification, both facing the weld. (Transducers preferably held in a fixture to control positioning—see sketch.) Amplitude calibration for pitch and catch is normally made by calibrating a single search unit. When switching to dual search units for pitch and catch inspection, there should be assurance that this calibration does not change as a result of instrument variables.

F—Further evaluate fusion zone indications with either 70 deg, 60 deg, or 45 deg transducer—whichever sound path is nearest to being perpendicular to the suspected fusion surface.

	Procedure legend		
	Area of weld thickness		
No.	Top quarter	Middle half	Bottom quarter
1	70°	70°	70°
2	60°	60°	60°
3	45°	45°	45°
4	60°	70°	70°
5	45°	70°	70°
6	70°G A	70°	60°
7	60° B	70°	60°
8	70°G A	60°	60°

	Procedure legend		
	Area of weld thickness		
No.	Top quarter	Middle half	Bottom quarter
9	70°G A	60°	45°
10	60° B	60°	60°
11	45° B	70°**	45°
12	70°G A	45°	70°G B
13	45° B	45°	45°
14	70°G A	45°	45°
15	70°G A	70°A B	70°G B

from the center of the transducer at one end to the center of the transducer at the other end.

6.19.5.7 Each weld discontinuity shall be accepted or rejected on the basis of its indication rating and its length, in accordance with Table 8.15.3 for buildings or Table 9.25.3 for bridges, whichever is applicable. Only those discontinuities which are rejectable need be recorded on the test report.

6.19.6 Each rejectable discontinuity shall be indicated on the weld by a mark directly over the discontinuity for its entire length. The depth from the surface and type of discontinuity shall be noted on nearby base metal.

6.19.7 Welds found unacceptable by ultrasonic testing shall be repaired by methods permitted by 3.7 of this code. Repaired welds shall be retested ultrasonically and an additional report form completed.

6.20 Preparation and Disposition of Reports

6.20.1 A report form which clearly identifies the work and the area of inspection shall be completed by the ultrasonic inspector at the time of inspection. The report form for welds which are acceptable need only contain sufficient information to identify the weld, the inspector (signature), and the acceptability of the weld. An example of such a form is shown in Appendix E.

6.20.2 Before a weld subject to ultrasonic testing by the contractor for the owner is accepted, all report forms pertaining to the weld, including any that show unacceptable quality prior to repair, shall be submitted to the Inspector.

6.20.3 A full set of completed report forms of welds subject to ultrasonic testing by the contractor for the owner, including any that show unacceptable quality prior to repair, shall be delivered to the owner upon completion of the work. The contractor's obligation to retain ultrasonic reports shall cease (1) upon delivery of this full set to the owner; (2) at the end of one full year after completion of the contractor's work, in the event that delivery is not required.

6.21 The Calibration of the Ultrasonic Unit with the IIW or Other Approved Calibration Blocks

(See Figs. 6.16.1A, 6.16.1B, and 6.21.)

6.21.1 Longitudinal Mode
  6.21.1.1 Distance Calibration
    (1) Set the transducer in position G on the IIW block, position H on the DC block, or position M on the DSC block.
    (2) Adjust instrument to produce indications at 1 in. (25.4 mm), 2 in. (50.8 mm), 3 in. (76.2 mm), 4 in. (102 mm), etc., on the CRT.
  6.21.1.2 Amplitude
    (1) Set the transducer in position G on the IIW block, position H on the DC block, or position M on the DSC block.
    (2) Adjust the gain until maximized indication from first back reflection attains 50 to 75% screen height.
  6.21.1.3 Resolution
    (1) Set the transducer in position F on the IIW block.
    (2) Transducer and instrument should resolve all three distances.

6.21.2 Shear Wave Mode (Transverse)
  6.21.2.1 Locate or check the transducer sound entry point (index point) by the following procedure:
    (1) Set the transducer in position D on the IIW block,

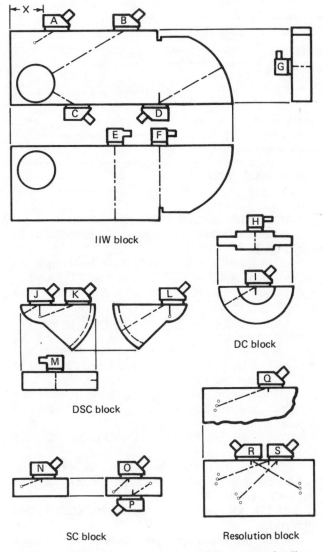

IIW block

DC block

DSC block

SC block

Resolution block

Fig. 6.21—Transducer positions (typical)

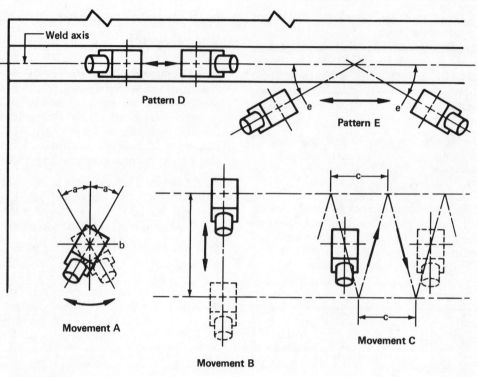

Notes:

1. Testing patterns are all symmetrical around the weld axis with the exception of pattern D which is conducted directly over the weld axis.

2. Testing from both sides of the weld axis is to be made wherever mechanically possible.

Fig. 6.22—Plan view of welded plate

position J or L on the DSC block, or I on the DC block.

(2) Move the transducer until the signal from the radius is maximized.

The point on the transducer which is in line with the line on the calibration block is indicative of the point of sound entry.

6.21.2.2 Check or determine the transducer sound path angle by the following procedure:

(1) Set the transducer in position B in IIW block for angles 40 deg through 60 deg.

(2) Set the transducer in position C on IIW block for angles 60 deg through 70 deg.

(3) Set the transducer in position K on DSC block for 45 deg through 70 deg.

(4) Set the transducer in position N on SC block for 70 deg angle.

(5) Set the transducer in position O on SC block for 45 deg angle.

(6) Set the transducer in position P on SC block for 60 deg angle.

(7) Move the transducer back and forth over the line indicative of the transducer angle until the signal from the radius is maximized. Compare the sound entry point on the transducer with the angle mark on the calibration block (Tolerance: ± 2 deg).

6.21.2.3 Distance Calibration Procedure

(1) Set the transducer in position D on the IIW block (any angle).

(2) Adjust the instrument to attain indications at 4 in. (102 mm) and 8 in. (204 mm) or 9 in. (230 mm) on the cathode ray tube (CRT), 9 in. on Type 1 block, or 8 in. on Type 2 block.

(3) Set the transducer in position J or L on the DSC block (any angle).

(4) Adjust the instrument to attain indications at 1 in. (25.4 mm), 5 in. (127 mm), and 9 in. (230 mm) on the CRT in the J position.

(5) Adjust the instrument to attain indications at 3 in. (76.2 mm) and 7 in. (177.8 mm) on the CRT in the L position.

(6) Set the transducer in position I on the DC block (any angle).

(7) Adjust the instrument to attain indication at 1 in. (25.4 mm), 2 in. (50.8 mm), 3 in. (76.2 mm), 4 in. (102 mm), etc., on the CRT.

6.21.2.4 Amplitude or Sensitivity Calibration Procedure

(1) Set the transducer in position A on the IIW block (any angle).

(2) Adjust the maximized signal from the 0.06 in.

(1.52 mm) hole to attain a horizontal reference line height indication.

(3) Set the transducer in position L on the DSC block (any angle).

(4) Adjust the maximized signal from the 1/32 in. (0.8 mm) slot to attain a horizontal reference line height indication.

(5) Set the transducer on the SC block, position N for 70 deg angle, position O for 45 deg angle, or position P for 60 deg angle.

(6) Adjust the maximized signal from the 1/16 in. (1.6 mm) hole to attain a horizontal reference line height indication.

(7) The decibel reading obtained in (6) shall be used as the "Reference Level" "b" reading on the Test Report sheet (Appendix E).

6.21.2.5 Resolution

(1) Set the transducer on resolution block, position Q for 70 deg angle, position R for 60 deg angle, or position S for 45 deg angle.

(2) Transducer and instrument shall resolve the three test holes, at least to the extent of distinguishing the peaks of the indications from the three holes.

6.21.2.6 Approach Distance of Search Unit

The minimum allowable distance, X, between the toe of the search unit and the edge of IIW block shall be as follows:

for 70 deg transducer, X = 2 in. (50.8 mm)
for 60 deg transducer, X = 1-5/8 in. (41.3 mm)
for 45 deg transducer, X = 1 in. (25.4 mm)

6.22 Scanning Patterns

(See Fig. 6.22)

6.22.1 Longitudinal Discontinuities

6.22.1.1 Scanning Movement A. Rotation angle a = 10 deg

6.22.1.2 Scanning Movement B. Scanning distance b shall be such that the section of weld being tested is covered.

6.22.1.3 Scanning Movement C. Progression distance c shall be approximately one-half the transducer width.

Note: Movements A, B, and C are combined into one scanning pattern.

6.22.2 Transverse Discontinuities

6.22.2.1 Scanning pattern D (when welds are ground flush).

6.22.2.2 Scanning pattern E (when weld reinforcement is not ground flush).

Scanning angle e = 15 deg max

Note: The scanning pattern is to be such that the full weld section is covered.

6.22.3 Electroslag or Electrogas Welds (Additional Scanning Pattern)—Scanning Pattern E

Search unit rotation angle e between 45 deg and 60 deg.

Note: The scanning pattern shall be such that the full weld section is covered.

7. Strengthening and Repairing Existing Structures

7.1 General

All provisions of this code apply equally to the strengthening and repairing of existing structures except as modified in this section.

7.2 Materials

7.2.1 The types of base metal involved shall be determined before preparing the drawings and specifications covering strengthening or repair of existing structures.

7.2.2 Where different base metals are to be joined, special consideration must be given to the selection of filler metal and welding procedure.

7.3 Design

7.3.1 Before completing the design, the following should be determined:
    7.3.1.1 The character and extent of damage to the parts and connections that require repair or strengthening.
    7.3.1.2 Whether the repairs should consist only of restoring corroded or otherwise damaged parts or of replacing members in entirety.

7.3.2 A complete study of stresses in the structure shall be made if the design of strengthening goes beyond the restoration of corroded or otherwise damaged members. Allowance should be made for fatigue stresses which members may have sustained in past service.

7.3.3 Members subject to cyclical loading shall be designed for fatigue stresses of the general specification. When the previous loading history is available for a structure, recognizance of it may be made to equalize the anticipated lives of old and replaced elements of the structure.

7.4 Workmanship

7.4.1 Surfaces of old material which are to be covered by repair or by reinforcing material shall be cleaned of dirt, rust, and other foreign matter except adherent paint film. The portions of such surfaces that are to receive welds shall be cleaned thoroughly of all foreign matter including paint film for a distance of 2 in. (50.8 mm) from each side of the outside lines of welds. Such surfaces, inside the areas cleaned for receiving welds, shall be given a protective coating if so specified.

7.4.2 Edges to be welded that have been reduced in thickness to less than the size of weld specified shall be cut away or built up to provide a thickness equal to the size of the weld except for occasional short lengths where some reduction of weld size would not be detrimental.

7.4.3 Structural elements under stress shall not be removed or reduced in section except as specified by the plans or Engineer.

7.5 Special

7.5.1 The Engineer shall determine whether or not a member is permitted to carry live-load stress while welding or oxygen cutting is being performed on it, taking into consideration the extent of cross-section heating of the member which results from the operation that is being performed.

7.5.2 If material is added to a member carrying a dead-load stress in excess of 3000 psi (20.7 MPa), it is desirable to relieve the member of dead-load stress or to prestress the material to be added. If neither is practicable, the new material to be added shall be proportioned for a unit stress equal to the allowable unit stress in the original member minus the dead-load unit stress in the original member.

7.5.3 Where rivets or bolts are overstressed by the total load, only dead load shall be assigned to them provided they are capable of supporting it without overstress. In such cases, sufficient welding shall be provided to support all live and impact loads. If rivets or bolts are overstressed by dead load alone, then sufficient welding shall be added to support the total load.

7.5.4 In strengthening members by the addition of material, it is desirable to arrange the sequence of welding so as to maintain a symmetrical section at all times. This is of particular importance if live load is permitted upon the structure while the member under consideration is being strengthened or repaired.

7.5.5 Particular care should be given to the sequence of welding in the application of reinforcing plates on girder webs and to the treatment of welds in the end joints of such plates where they abut stiffener assemblies or girder splice plates.

8. Design of New Buildings

Part A
General Requirements

8.1 Application

8.1.1 This section supplements Sections 1 through 6 and is to be used in conjunction with the prescribed Building Code[32] for the design and construction of steel structures.

8.1.2 Where fatigue loading would govern the proportions of a member or its connection, the provisions of Appendix B of the AISC Specification for the Design, Fabrication and Erection of Structural Steel for Buildings should take precedence over the values tabulated in this section.

8.2 Base Metal

8.2.1 Steel base metal to be welded under this code shall conform to the requirements of the latest edition of one of the specifications listed below. Combinations of any of the steel base metals specified may be welded together.

8.2.1.1 ASTM A36, Specification for Structural Steel.

8.2.1.2 ASTM A53, Grade B, Specification for Pipe, Steel, Black and Hot-Dipped, Zinc-Coated, Welded and Seamless.

8.2.1.3 ASTM A242, Specification for High-Strength Low-Alloy Structural Steel (if the properties are suitable for welding).

8.2.1.4 ASTM A441, Specification for High-Strength Low-Alloy Structural Manganese-Vanadium Steel.

8.2.1.5 ASTM A500, Specification for Cold-Formed Welded and Seamless Carbon Steel Structural Tubing in Rounds and Shapes.

8.2.1.6 ASTM A501, Specification for Hot-Formed Welded and Seamless Carbon Steel Structural Tubing.

8.2.1.7 ASTM A514, Specification for High-Yield-Strength, Quenched and Tempered Alloy Steel Plate, Suitable for Welding.

8.2.1.8 ASTM A516, Specification for Pressure Vessel Plates, Carbon Steel, for Moderate- and Lower-Temperature Service.

8.2.1.9 ASTM A517, Specification for Pressure Vessel Plates, Alloy Steel, High-Strength, Quenched and Tempered.

8.2.1.10 ASTM A529, Specification for Standard Steel with 42 000 psi Minimum Yield Point (1/2 in. Maximum Thickness).

8.2.1.11 ASTM A570, Specification for Hot-Rolled Carbon Steel Sheet and Strip, Structural Quality.

8.2.1.12 ASTM A572, Specification for High-Strength Low-Alloy Columbium-Vanadium Steels of Structural Quality.

8.2.1.13 ASTM A588, Specification for High-Strength Low-Alloy Structural Steel with 50 000 psi Minimum Yield Point to 4 in. Thick.

8.2.1.14 ASTM A606, Type 2 (Type 4 if the properties are suitable for welding), Specification for Steel Sheet and Strip, Hot-Rolled and Cold-Rolled, High Strength, Low Alloy, with Improved Corrosion Resistance.

8.2.1.15 ASTM A607, Grades 45, 50, and 55, Specification for Steel Sheet and Strip, Hot-Rolled or Cold-Rolled, High-Strength, Low-Alloy Columbium and/or Vanadium.

8.2.1.16 ASTM A618, Grades II and III (Grade I if the properties are suitable for welding), Specification for Hot-Formed Welded and Seamless High-Strength, Low-Alloy Structural Tubing.

8.2.1.17 ASTM A633, Specification for Normalized High-Strength Low-Alloy Structural Steel.

8.2.1.18 ASTM A709, Specification for Structural Steel for Bridges.

8.2.2 When an ASTM A709 grade of structural steel is considered for use, its weldability shall be established by

---

32. The term "Building Code," whenever the expression occurs in this code, refers to the building law or specifications or other construction regulations in conjunction with which this code is applied. In the absence of any locally applicable building law or specifications or other construction regulations, it is recommended that the construction be required to comply with the Specification for the Design, Fabrication and Erection of Structural Steel for Buildings of the American Institute of Steel Construction (AISC).

Table 8.4.1
Allowable stresses in welds

Type of weld	Stress in weld[1]	Allowable stress	Required weld strength level[2]
Complete joint penetration groove welds	Tension normal to the effective area	Same as base metal	Matching weld metal must be used. See Table 4.1.1.
	Compression normal to the effective area	Same as base metal	Weld metal with a strength level equal to or one classification (10 ksi) less than matching weld metal may be used.
	Tension or compression parallel to the axis of the weld	Same as base metal	
	Shear on the effective area	0.30 nominal tensile strength of weld metal (ksi), except shear stress on base metal shall not exceed 0.40 yield strength of base metal	Weld metal with a strength level equal to or less than matching weld metal may be used.

1. For definition of effective area, see 2.3.
2. For matching weld metal, see Table 4.1.1.
3. Fillet welds and partial joint penetration groove welds joining the component elements of built-up members, such as flange-to-web connections, may be designed without regard to the tensile or compressive stress in these elements parallel to the axis of the welds.

the steel producer, and the procedure for welding it shall be established by qualification in accordance with the requirements of 5.2 and other such requirements as may be prescribed by the Engineer, with the following exception: If the grade supplied meets the chemical and mechanical properties of ASTM A36, A572 Gr. 50, A588, or A514, the applicable prequalified procedures of this code shall apply.

8.2.3 When a steel other than those listed in 8.2.1 is approved under the provisions of the general building code and such steel is proposed for welded construction, the weldability of the steel and the procedure for welding it shall be established by qualification in accordance with the requirements of 5.2 and such other requirements as prescribed by the Engineer.

The responsibility for determining weldability, including the assumption of the additional testing cost involved, is assigned to the party who either specifies a material not listed in 8.2.1 or who proposes the use of a substitute material not listed in 8.2.1. The fabricator shall have the responsibility of establishing the welding procedure by qualification.

8.2.4 Extension bars and runoff plates used in welding shall conform to the following requirements:

(1) When used in welding with an approved steel listed in 8.2.1, they may be any of the steels listed in 8.2.1.

(2) When used in welding with a steel qualified in accordance with 8.2.3 they may be

(a) The steel qualified, or

(b) Any steel listed in 8.2.1

Steel for backing shall conform to the requirements of (1) and (2), except that 100 ksi minimum yield strength steel as backing shall only be used with 100 ksi minimum yield strength steel.

Spacers used shall be of the same material as the base metal.

8.2.5 The provisions of this code are not intended for use with steels having a minimum specified yield point or yield strength over 100 000 psi (690 MPa).

Part B
Allowable Unit Stresses

8.3 Base Metal Stresses

The base steel stresses shall be those specified in the applicable Building Code.

Table 8.4.1 (continued)
Allowable stresses in welds

Type of weld	Stress in weld[1]		Allowable stress	Required weld strength level[2]
Partial joint penetration groove welds	Compression normal to effective area	Joint not designed to bear	0.50 nominal tensile strength of weld metal (ksi), except stress on base metal shall not exceed 0.60 yield strength of base metal	Weld metal with a strength level equal to or less than matching weld metal may be used.
		Joint designed to bear	Same as base metal	
	Tension or compression parallel to the axis of the weld[3]		Same as base metal	
	Shear parallel to axis of weld		0.30 nominal tensile strength of weld metal (ksi), except shear stress on base metal shall not exceed 0.40 yield strength of base metal	
	Tension normal to effective area		0.30 nominal tensile strength of weld metal (ksi), except shear stress on base metal shall not exceed 0.55 yield strength of base metal	
Fillet welds	Shear on effective area		0.30 nominal tensile strength of weld metal (ksi), except shear stress on base metal shall not exceed 0.40 yield strength of base metal	Weld metal with a strength level equal to or less than matching weld metal may be used.
	Tension or compression parallel to axis of weld[3]		Same as base metal	
Plug and slot welds	Shear parallel to faying surfaces (on effective area)		0.30 nominal tensile strength of weld metal (ksi), except shear stress on base metal shall not exceed 0.40 yield strength of base metal	Weld metal with a strength level equal to or less than matching weld metal may be used.

1. For definition of effective area, see 2.3.
2. For matching weld metal, see Table 4.1.1.
3. Fillet welds and partial joint penetration groove welds joining the component elements of built-up members, such as flange-to-web connections, may be designed without regard to the tensile or compressive stress in these elements parallel to the axis of the welds.

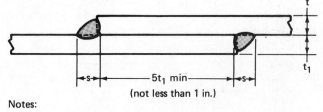

Notes:

1. s = as required
2. $t > t_1$

Fig. 8.8.3—Double-fillet-welded lap joint

8.4 Unit Stresses in Welds

8.4.1 Except as modified by 8.5, allowable unit stresses in welds shall not exceed those listed in Table 8.4.1.

8.4.2 Stress on the effective throat of fillet welds is considered as shear stress regardless of the direction of application.

8.5 Increased Unit Stresses

Where the Building Code permits the use of increased unit stresses in the base metal for any reason, a corresponding increase shall be applied to the allowable unit stress for welds.

Part C
Structural Details

8.6 Combinations of Welds

If two or more of the general types of welds (groove, fillet, plug, slot) are combined in a single joint, their allowable capacity shall be computed with reference to the axis of the group in order to determine the allowable capacity of the combination. See Appendix A. However, such methods of adding individual capacities of welds does not apply to fillet welds reinforcing groove welds.

8.7 Welds in Combination with Rivets and Bolts

Rivets or bolts used in bearing-type connections shall not be considered as sharing the stress in combination with welds. Welds, if used, shall be provided to carry the entire stress in the connection. However, connections that are welded to one member and riveted or bolted to the other member are permitted. High strength bolts properly installed as a friction-type connection prior to welding may be considered as sharing the stress with the welds.

8.8 Fillet Weld Details

8.8.1 If longitudinal fillet welds are used alone in end connections of flat bar tension members, the length of each fillet weld shall be no less than the perpendicular distance between them. The transverse spacing of longitudinal fillet welds used in end connections shall not exceed 8 in. (203 mm) unless end transverse welds or intermediate plug or slot welds are used.

8.8.2 Intermittent fillet welds may be used to carry calculated stress.

8.8.3 For lap joints, the minimum amount of lap shall be five times the thickness of the thinner part joined but not less than 1 in. (25.4 mm) (see Fig. 8.8.3).

8.8.4 Lap joints in parts carrying axial stress shall be double-fillet welded (see Fig. 8.8.3) except where deflection of the joint is sufficiently restrained to prevent it from opening under load.

8.8.5 Fillet welds deposited on the opposite sides of a common plane of contact between two parts shall be interrupted at a corner common to both welds. See Fig. 8.8.5.

8.8.6 Boxing (End Returns)

8.8.6.1 Side or end fillet welds terminating at ends or sides, respectively, of parts or members shall, wherever practicable, be returned continuously around the corners for a distance at least twice the nominal size of the weld except as provided in 8.8.5. This provision shall apply to side and top fillet welds connecting brackets, beam seats, and similar connections on the plane about which bending moments are computed.

8.8.6.2 End returns shall be indicated on the drawings.

8.9 Eccentricity

In general, adequate provision shall be made for bending stresses due to eccentricity, if any, in the disposition and section of base metal parts and in the location and types of welded joints. The disposition of fillet welds to balance the forces about the neutral axis or axes for end connections of single-angle, double-angle, and similar type members is not required; such weld arrangements at the heel and toe of angle members may be distributed to conform to the length of the various available edges. Similarly, T's or beams framing into chords of trusses, or similar joints, may be connected with unbalanced fillet welds.

8.10 Transition of Thicknesses or Widths

Tension butt joints in axially aligned primary members or different material thicknesses or widths shall be made in such a manner that the slope through the transition zone does not exceed 1 in 2-1/2. The transition shall be accomplished by chamfering the thicker part, tapering the wider part, sloping the weld metal, or by any combination of these (see Fig. 8.10).

8.11 Beam End Connections

Welded beam end connections shall be designed in accordance with the assumptions about the degree of restraint involved in the designated type of construction.

8.12 Connections of Components of Built-Up Members

8.12.1 If two or more plates or rolled shapes are used to build up a member, sufficient stitch welding (of the fillet, plug, or slot type) shall be provided to make the parts act in unison as follows, except where transfer or calculated stress between the parts joined requires closer spacing.

8.12.1.1 The maximum longitudinal spacing of stitch welds connecting two or more rolled shapes in contact with one another shall not exceed 24 in. (610 mm).

8.12.1.2 In built-up compression members, the longitudinal spacing of stitch welds connecting a plate component to other components shall not exceed the plate thickness times $4000/\sqrt{F_y}$ nor shall it exceed 12 in. (305 mm) (F_y = specified minimum yield point in psi of the type of steel being used). The unsupported width of web, cover, or diaphragm plates, between adjacent lines of welds, shall not exceed the plate thickness times $8000/\sqrt{F_y}$. When the unsupported width exceeds this limit, but a portion of its width no greater than $8000/\sqrt{F_y}$ times the thickness would satisfy the stress requirements, the member will be considered acceptable.

8.12.1.3 In built-in tension members, the longitudinal spacing of stitch welds connecting a plate component to other components, or connecting two plate components to each other, shall not exceed 12 in. (305 mm) or 24 times the thickness of the thinner plate.

Part D
Workmanship

8.13 Dimensional Tolerances

The dimensions of structural members shall be within the tolerances specified in 3.5, with the following additional requirements.

8.13.1 Variation from flatness of girder webs shall be determined by measuring offsets from a straight edge whose length is not less than the least dimension of any panel. The straight edge shall be placed in any position of maximum variation on the web with the ends of the straight edge adjacent to opposite panel boundaries.

8.13.2 Permissible variations from flatness of webs having a depth, D, and a thickness, t, in panels bounded by

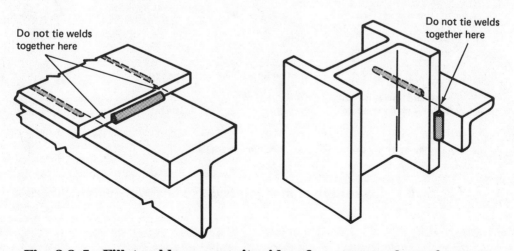

Fig. 8.8.5—Fillet welds on opposite sides of a common plane of contact

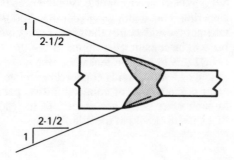
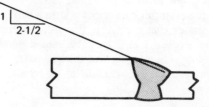

Transition by sloping weld surface

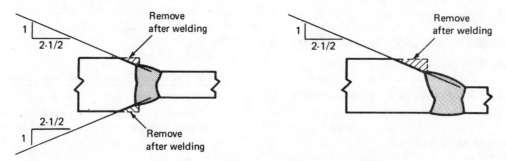

Transition by sloping weld surface and chamfering

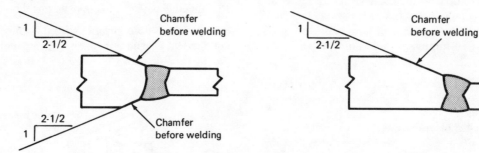

Transition by chamfering thicker part

Center line alignment
(particularly applicable to web plates)

Offset alignment
(particularly applicable to flange plates)

Notes:

1. Groove may be of any permitted or qualified type and detail.

2. Transition slopes shown are the maximum permitted.

Fig. 8.10A—Transition of butt joints in parts of unequal thickness

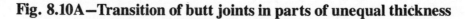

stiffeners or flanges, or both, whose least panel dimension is *d* shall not exceed the following:

	Loading	
	Dynamic	Static
Intermediate stiffeners on both sides of web		
where $D/t < 150$, maximum variation = $d/115$		$d/100$
where $D/t \geq 150$, maximum variation = $d/92$		$d/80$
Intermediate stiffeners on one side only of web		
where $D/t < 100$, maximum variation = $d/100$		
where $D/t \geq 100$, maximum variation = $d/67$		
No intermediate stiffeners		
maximum variation = $D/150$		

(See Appendix G for tabulation.)

8.13.3 Web distortions of twice the allowable tolerances of 8.13.1.2 shall be satisfactory when occurring at the end of a girder which has been drilled, or subpunched and reamed; either during assembly or to a template for a field bolted splice; provided, when the splice plates are bolted, the web assumes the proper dimensional tolerances.

8.13.4 If architectural considerations require tolerances more restrictive than described in 8.13.1, specific reference must be included in the bid documents.

8.14 Temporary Welds

Temporary welds shall be subject to the same welding procedure requirements as final welds. They shall be removed when required by the Engineer. When they are removed, the surface shall be made flush with the original surface.

8.15 Quality of Welds

8.15.1 Visual Inspection. All welds shall be visually inspected. A weld shall be acceptable by visual inspection if it shows that

8.15.1.1 The weld has no cracks.

8.15.1.2 Thorough fusion exists between adjacent layers of weld metal and between weld metal and base metal.

8.15.1.3 All craters are filled to the full cross section of the weld.

8.15.1.4 Weld profiles are in accordance with 3.6.

8.15.1.5 The sum of diameters of piping porosity in fillet welds does not exceed 3/8 in. (9.5 mm) in any linear inch of weld and shall not exceed 3/4 in. (19.0 mm) in any 12 in. (305 mm) length of weld.

8.15.1.6 A fillet weld in any single continuous weld shall be permitted to underrun the nominal fillet size required by 1/16 in. (1.6 mm) without correction, provided that the undersize portion of the weld does not exceed 10% of the length of the weld. On web-to-flange welds on girders, no underrun is permitted at the ends for a length equal to twice the width of the flange.

8.15.1.7 Complete joint penetration groove welds in butt joints transverse to the direction of computed tensile stress shall have no piping porosity. For all other groove welds, piping porosity shall not exceed 3/8 in. (9.5 mm) in any linear inch of weld and shall not exceed 3/4 in. (19 mm) in any 12 in. (305 mm) length of weld.

8.15.1.8 Visual inspection of welds in all steels may begin immediately after the completed welds have cooled to ambient temperature. Acceptance criteria for ASTM A514 and A517 steels shall be based on visual inspection performed not less than 48 hours after completion of the weld.

8.15.2 Radiographic and Magnetic Particle Inspection. Welds that are subject to radiographic or magnetic particle testing in addition to visual inspection shall have no cracks and shall be unacceptable if the radiograph or magnetic particle inspection shows any of the types of discontinuities given in 8.15.2.1 or 8.15.2.2.

8.15.2.1 Individual discontinuities, having a greatest dimension of 3/32 in. (2.4 mm) or greater, if

(1) The greatest dimension of a discontinuity is larger than 2/3 of the effective throat, 2/3 the weld size, or 3/4 in. (19.0 mm).

(2) The discontinuity is closer than three times its greatest dimension to the end of a groove weld subject to primary tensile stresses.

(3) A group of such discontinuities is in line such that

(a) The sum of the greatest dimensions of all such discontinuities is larger than the effective throat or weld size in any length of six times the effective throat or weld size. When the length of the weld being examined is less than six times the effective throat or weld size, the permissible sum of the greatest dimensions shall be proportionally less than the effective throat or weld size.

(b) The space between two such discontinuities which are adjacent is less than three times the greatest dimension of the larger of the discontinuities in the pair being considered.

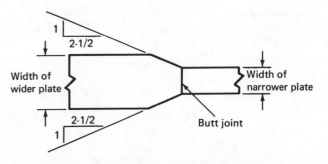

Fig. 8.10B—Transition of thicknesses or widths

8.15.2.2 Independent of the requirements of 8.15.2.1, discontinuities having a greatest dimension of less than 3/32 in. (2.4 mm), if the sum of their greatest dimensions exceeds 3/8 in. (9.5 mm) in any linear inch of weld.

8.15.3 Ultrasonic Inspection. Welds that are subject to ultrasonic testing, in addition to visual inspection, shall be acceptable if they meet the requirements of Table 8.15.3. Ultrasonically tested welds are evaluated on the basis of a discontinuity reflecting ultrasound in proportion to its effect on the integrity of the weld.

8.15.4 Liquid Penetrant Inspection. Welds that are subject to liquid penetrant testing, in addition to visual inspection, shall be evaluated on the basis of the requirements for visual inspection.

8.15.5 When welds are subject to nondestructive testing in accordance with 8.15.2, 8.15.3, and 8.15.4, the testing may begin immediately after the completed welds have cooled to ambient temperature. Acceptance criteria for ASTM A514 and A517 steels shall be based on nondestructive testing performed not less than 48 hours after completion of the welds.

Table 8.15.3
Ultrasonic acceptance criteria

	Minimum acceptance levels (decibels)															
	Weld thickness (in.) and transducer angle															
Reflector severity	5/16 to 3/4	>3/4 to 1-1/2	>1-1/2 to 2-1/2			>2-1/2 to 4			>4 to 6			>6 to 8		
	70°	70°	70°	60°	45°	70°	60°	45°	70°	60°	45°	70°	60°	45°
Large reflectors	+ 8	+ 3	− 1	+ 2	+ 4	− 4	− 1	+ 1	− 7	− 4	− 2	− 9	− 6	− 4
Small reflectors	+ 9	+ 4	+ 1	+ 4	+ 6	− 2	+ 1	+ 3	− 5	− 2	0	− 7	− 4	− 2
Minor reflectors	+10	+ 5	+ 3	+ 6	+ 8	0	+ 3	+ 5	− 3	0	+ 2	− 5	− 2	0

Notes:
1. Discontinuities which have a more serious rating than those of minor reflectors shall be separated by at least 2L, L being the length of the larger discontinuity. Discontinuities not separated by at least 2L are considered to be one discontinuity whose length is determined by the combined length of the discontinuities plus their separation distance.
2. Discontinuities which have a more serious rating than those of minor reflectors shall not begin at a distance smaller than 2L from weld ends carrying primary tensile stress, L being the discontinuity length.
3. Discontinuities detected at 'scanning levels' in the root-face areas of complete joint penetration double-V-groove welds, double-J-groove welds, double-U-groove welds, and double-bevel-groove welds shall be evaluated at an acceptance level 4 db* more sensitive than prescribed in this table when such welds are designated on design drawings as 'tension welds.'
4. Electroslag and electrogas welds—discontinuities which exceed 2 in. (51 mm) in length and occur in the middle half of such welds are to be evaluated at an acceptance level 6 db more sensitive than the above levels.

*i.e., add +4 dB to the number in the table.

Large reflectors
Any discontinuity, REGARDLESS OF LENGTH, having a more serious rating (smaller number) than this level shall be rejected.

Small reflectors
Any discontinuity longer than 3/4 in. (19.0 mm) having a more serious rating (smaller number) than this level shall be rejected.

Minor reflectors
Only those discontinuities exceeding 2 in. (51 mm) in length and having a more serious rating (smaller number) than this level shall be rejected.

Scanning levels		
Sound path distance		Above zero reference, dB
in.	mm	
To 2-1/2	63.5	+14
>2-1/2 to 5	63.5 - 127	+19
>5 to 10	127 - 254	+29
>10 to 15	254 - 381	+39

9. Design of New Bridges

Part A
General Requirements

9.1 Application

This section supplements Sections 1 through 6 and is to be used in conjunction with the prescribed standard specification for the design and construction of highway or railroad bridges, as required.

9.2 Base Metal

9.2.1 Steel base metal to be welded under this code shall conform to the requirements of the latest edition of one of the specifications listed below. Combinations of any approved steel base metals may be welded together.

9.2.1.1 ASTM A36, Specification for Structural Steel.

9.2.1.2 ASTM A441, Specification for High-Strength Low-Alloy Structural Manganese-Vanadium Steel.

9.2.1.3 ASTM A500, Specification for Cold-Formed Welded and Seamless Carbon Steel Structural Tubing in Rounds and Shapes.

9.2.1.4 ASTM A501, Specification for Hot-Formed Welded and Seamless Carbon Steel Structural Tubing.

9.2.1.5 ASTM A514, Specification for High-Yield-Strength, Quenched and Tempered Alloy Steel Plate, Suitable for Welding.

9.2.1.6 ASTM A516, Specification for Pressure Vessel Plates, Carbon Steel, for Moderate- and Lower-Temperature Service.

9.2.1.7 ASTM A572, Grades 42 and 50, Specification for High-Strength Low-Alloy Columbium-Vanadium Steels of Structural Quality.

9.2.1.8 ASTM A588, Specification for High-Strength Low-Alloy Structural Steel with 50 000 psi Minimum Yield Point to 4 in. Thick.

9.2.1.9 ASTM A618, Grades II and III (Grade I if the properties are suitable for welding), Specification for Hot-Formed Welded and Seamless High-Strength Low-Alloy Structural Tubing.

9.2.1.10 ASTM A633, Specification for Normalized High-Strength Low-Alloy Structural Steel.

9.2.1.11 ASTM A709, Specification for Structural Steel for Bridges.

9.2.2 When an ASTM A709 Grade structural steel is considered for use, its weldability shall be established by the steel producer and the procedure for welding it shall be established by qualification in accordance with the requirements of 5.2 and such other requirements as may be prescribed by the Engineer, with the following exception: If the grade to be supplied meets the chemical and mechanical properties of ASTM A36, A572 Gr. 50, A588, or A514, the applicable prequalified procedures of this code shall apply.

9.2.3 When an ASTM A242 or A618, Grade 1, low-alloy steel is considered for use, its weldability shall be investigated by the Engineer, and the Engineer shall specify all pertinent material, design, and workmanship information not covered by this code.

9.2.4 When a steel other than one listed in 9.2.1 is approved under the provisions of the general bridge specifications and such steel is proposed for welded construction, the weldability of the steel and the procedure for welding it shall be established by qualification in accordance with the requirements of 5.2 and such other requirements as prescribed by the Engineer.

9.2.4.1 The responsibility for determining weldability, including the assumption of the additional testing costs involved, is assigned to the party who either specifies a material not listed in 9.2.1 or who proposes the use of a substitute material not listed in 9.2.1. The fabricator shall have the responsibility of establishing the welding procedure by qualification.

9.2.5 Extension bars and runoff plates used in welding shall conform to the following requirements:

(1) When used in welding with an approved steel listed in 9.2.1, they may be any of the steels listed in 9.2.1.

Table 9.3.1
Allowable stresses in welds

Type of weld	Stress in weld[1]	Allowable stress	Required weld strength level[2]
Complete joint penetration groove welds	Tension normal to the effective area	Same as base metal	Matching weld metal must be used. See Table 4.1.1.
	Compression normal to the effective area	Same as base metal	Weld metal with a strength level equal to or one classification (10 ksi) less than matching weld metal may be used.
	Tension or compression parallel to the axis of the weld	Same as base metal	Weld metal with a strength level equal to or less than matching weld metal may be used.
	Shear on the effective area	0.27 nominal tensile strength of weld metal (ksi), except shear stress on base metal shall not exceed 0.36 yield strength of base metal	

1. For definition of effective area, see 2.3.
2. For matching weld metal, see Table 4.1.1.
3. Fillet welds and partial joint penetration groove welds joining the component elements of built-up members, such as flange-to-web connections, may be designed without regard to the tensile or compressive stress in these elements parallel to the axis of the welds.

(2) When used in welding with a steel qualified in accordance with 9.2.4, they may be

    (a) The steel qualified, or

    (b) Any steel listed in 9.2.1

Steel for backing shall conform to the requirements of (1) and (2), except that 100 ksi minimum yield strength steel as backing shall only be used with 100 ksi minimum yield strength steel.

Spacers shall be of the same material as the base metal.

9.2.6 The provisions of this code are not intended for use with steels having a minimum specified yield point or yield strength over 100 000 psi (690 MPa).

Part B
Allowable Unit Stresses

9.3 Unit Stresses in Welds[33]

Note: The application of these stresses is modified by the requirements of 9.4.

---

33. Unless specified in the general specifications, it is recommended that the basic unit shear stress in the next section be 65% of the basic allowable stress in tension.

9.3.1 Except as modified by 9.4, 9.5, and 9.6, allowable unit stresses in welds shall not exceed those listed in Table 9.3.1.

9.3.2 Stress on the effective throat of fillet welds is considered as shear stress regardless of the direction of application.

9.4 Fatigue Stress Provisions

The fatigue stress provisions, as applicable, comply with the Standard Specifications for Highway Bridges as adopted by the American Association of State Highway and Transportation Officials (AASHTO), or Specification for Steel Railway Engineering Association (AREA). For bridges subject to cyclic loading, other than highway or railway applications, stress ranges may be obtained from Table 9.4 and Figs. 9.4A, 9.4B, and 9.4C for appropriate general condition and cycle life. The cycle life should be determined by the Engineer to meet the planned life requirements of the structure.

Table 9.3.1 (continued)
Allowable stresses in welds

Type of weld	Stress in weld[1]		Allowable stress	Required weld strength level[2]
Partial-joint penetration groove welds	Compression normal to effective area	Joint not designed to bear	0.45 nominal tensile strength of weld metal (ksi), except stress on base metal shall not exceed 0.55 yield strength of base metal	Weld metal with a strength level equal to or less than matching weld metal may be used.
		Joint designed to bear	Same as base metal	
	Tension or compression parallel to the axis of the weld[3]		Same as base metal	
	Shear parallel to axis of weld		0.27 nominal tensile strength of weld metal (ksi), except shear stress on base metal shall not exceed 0.36 yield strength of base metal	
	Tension normal to effective area		0.27 nominal tensile strength of weld metal (ksi), except shear stress on base metal shall not exceed 0.60 yield strength of base metal	
Fillet welds	Shear on effective area		0.27 nominal tensile strength of weld metal (ksi), except shear stress on base metal shall not exceed 0.36 yield strength of base metal	Weld metal with a strength level equal to or less than matching weld metal may be used.
	Tension or compression parallel to axis of weld[3]		Same as base metal	
Plug and slot welds	Shear parallel to faying surfaces (on effective area)		0.27 nominal tensile strength of weld metal (ksi), except shear stress on base metal shall not exceed 0.36 yield strength of base metal	Weld metal with a strength level equal to or less than matching weld metal may be used.

1. For definition of effective area, see 2.3.
2. For matching weld metal, see Table 4.1.1.
3. Fillet welds and partial joint penetration groove welds joining the component elements of built-up members, such as flange-to-web connections, may be designed without regard to the tensile or compressive stress in these elements parallel to the axis of the welds.

Table 9.4
Fatigue stress provisions—tension or reversal stresses*

General condition	Situation	Stress category (see Fig. 9.4A)	Example (see Fig. 9.4A)
Plain material	Base metal with rolled or cleaned surfaces. Oxygen-cut edges with ANSI smoothness of 1000 or less.	A	1, 2
Built-up members	Base metal and weld metal in members without attachments, built up of plates or shapes connected by continuous complete or partial joint penetration groove welds or by continuous fillet welds parallel to the direction of applied stress.	B	3, 4, 5, 7
	Calculated flexural stress at toe of transverse stiffener welds on girder webs or flanges.	C	6
	Base metal at end of partial length welded cover plates having square or tapered ends, with or without welds across the ends.	E	7
Groove welds	Base metal and weld metal at complete joint penetration groove welded splices of rolled and welded sections having similar profiles when welds are ground[1] and weld soundness established by nondestructive testing.[2]	B	8, 9
	Base metal and weld metal in or adjacent to complete joint penetration groove welded splices at transitions in width or thickness, with welds ground[1] to provide slopes no steeper than 1 to 2-1/2[3] and weld soundness established by nondestructive testing.[2]	B	10, 11

Groove welded connections	Base metal at details of any length attached by groove welds subjected to transverse or longitudinal loading, or both, when weld soundness is transverse to the direction of stress is established by nondestructive testing[2] and the detail embodies a transition radius, R, equal to or greater than 2 in. with the weld termination ground[1] when	Longitudinal loading	Transverse loading[4]			Example (see Fig. 9.4A)
			Materials having equal or unequal thickness sloped,[6] welds ground,[1] web connections excluded.	Materials having equal thickness, not ground; web connections excluded.	Materials having unequal thickness, not sloped or ground, including web connections.	
	(a) R ≥ 24 in.	B	B	C	E	13
	(b) 24 in. > R ≥ 6 in.	C	C	C	E	13
	(c) 6 in. ≥ R ≥ 2 in.	D	D	D	E	13

*Except a noted for fillet and stud welds.

Table 9.4 (continued)
Fatigue stress provisions—tension or reversal stresses*

General condition	Situation	Stress category (see Fig. 9.4A)	Example (see Fig. 9.4A)
Groove welds	Base metal and weld metal in or adjacent to complete joint penetration groove welded splices either not requiring transition or when required with transitions having slopes no greater than 1 to 2-1/2[3] and when in either case reinforcement is not removed and weld soundness is established by nondestructive testing.[2]	C	8, 9, 10, 11
Groove or fillet welded connections	Base metal at details attached by groove or fillet welds subject to longitudinal loading when the detail length, L, parallel to the line of stress is		
	(a) < 2 in.	C	12,14,15,16
	(b) 2 in. ≤ L<4 in.	D	12
	(c) L ≥ 4 in.	E	12
Fillet welded connections	Base metal at details attached by fillet welds parallel to the direction of stress regardless of length when the detail embodies a transition radius, R, 2 in. or greater and with the weld termination ground.[1]		
	(a) When R ≥ 24 in.	B[5]	13
	(b) When 24 in. > R ≥ 6 in.	C[5]	13
	(c) When 6 in. > R≥ 2 in.	D[5]	13
Fillet welds	Shear stress on throat of fillet welds.	F	8a
	Base metal at intermittent welds attaching transverse stiffeners and stud-type shear connectors.	C	7,14
	Base metal at intermittent fillet welds attaching longitudinal stiffeners.	E	—
Stud welds	Shear stress on nominal shear area of stud-type shear connectors.	F	14
Plug and slot welds	Base metal adjacent to or connected by plug or slot welds.	E	—

1. Finished according to 3.6.3.
2. Either RT or UT to meet quality requirements of 9.25.2 or 9.25.3 for welds subject to tensile stress.
3. Sloped as required by 9.20.1.
4. Applicable only to complete joint penetration groove welds.
5. Shear stress on throat of weld (loading through the weld in any direction) is governed by Category F.
6. Slopes similar to those required by Footnote 3 are mandatory for categories listed. If slopes are not obtainable, Category E must be used.

*Except as noted for fillet and stud welds.

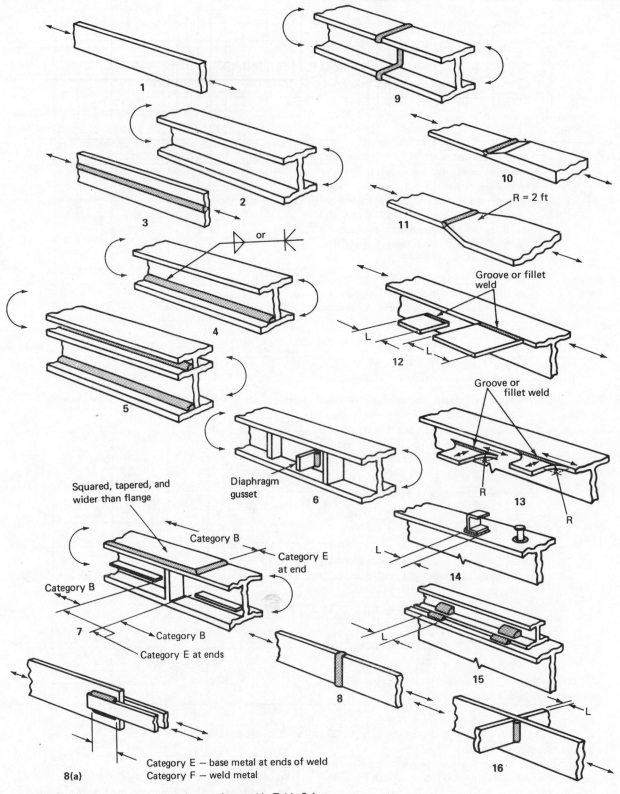

Squared, tapered, and wider than flange

Diaphragm gusset

Category B

Category E at end

Category B

Category B

Category E at ends

Groove or fillet weld

Groove or fillet weld

R = 2 ft

Category E — base metal at ends of weld
Category F — weld metal

Note: The numbers below each example are referenced in Table 9.4

Fig. 9.4A—Examples of various fatigue categories

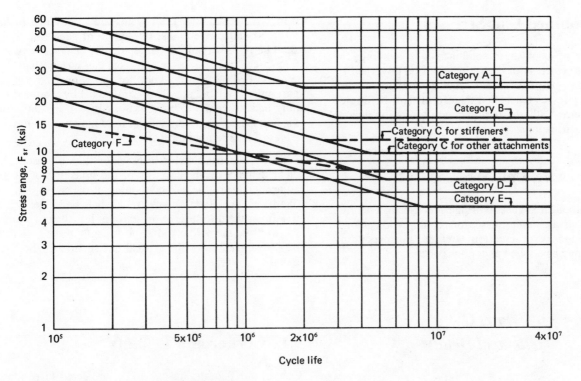

*Transverse stiffener welds on girder webs or flanges

Fig. 9.4B—Design stress range curves for categories A to F—redundant structures

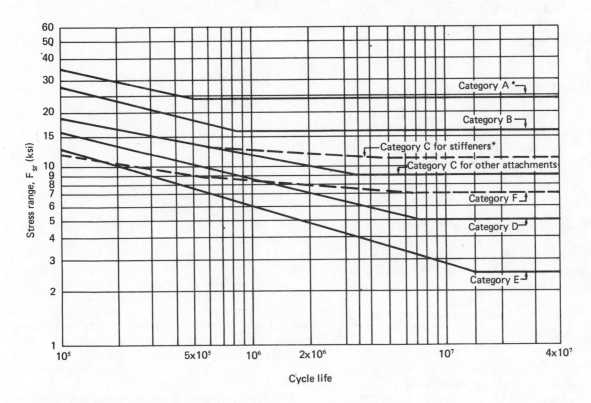

*Transverse stiffener welds on girder webs or flanges

Fig. 9.4C—Design stress range curves for categories A to F—nonredundant structures

9.5 Combined Unit Stresses

In the case of axial stress combined with bending, the allowable unit stress of each kind shall be governed by the requirements of 9.3 and 9.4 and the maximum combined unit stresses calculated therefrom shall be limited in accordance with the requirements of the applicable general specifications.

9.6 Increased Unit Stresses

When the applicable general bridge specification permits the use of increased unit stresses for a combination of loads or for secondary or erection stresses, corresponding increases may be applied under this code.

Part C
Structural Details

9.7 General

In general, details shall minimize constraint against ductile behavior, avoid undue concentration of welding, and afford ample access for depositing the weld metal.

9.8 Noncontinuous Beams

The connections at the ends of noncontinuous beams shall be designed with flexibility so as to avoid excessive secondary stresses due to bending. Seated connections with a flexible or guiding device to prevent end twisting are recommended.

9.9 Participation of Floor System

Details of the floor system should be so designed as to avoid, insofar as possible, unintended participation in the chord or flange stresses.

9.10 Lap Joints

9.10.1 The minimum overlap of parts in stress-carrying lap joints shall be five times the thickness of the thinner part. Unless lateral deflection of the parts is prevented, they shall be connected by at least two transverse lines of fillet, plug, or slot welds or by two or more longitudinal fillet or slot welds.

9.10.2 If longitudinal fillet welds are used alone in lap joints of end connections, the length of each fillet weld shall be no less than the perpendicular distance between them. The transverse spacing of the welds shall not exceed 16 times the thickness of the connected thinner part unless suitable provision is made (as by intermediate plug or slot welds) to prevent buckling or separation of the parts. The longitudinal fillet weld may be either at the edges of the member or in slots.

9.10.3 When fillet welds in holes or slots are used, the clear distance from the edge of the hole or slot to the adjacent edge of the part containing it, measured perpendicular to the direction of stress, shall be no less than five times the thickness of the part nor less than two times the width of the hole or slot. The strength of the part shall be determined from the critical net section of the base metal.

9.11 Corner and T-Joints

Corner and T-joints that are to be subjected to bending about an axis parallel to the joint shall have their welds arranged to avoid concentration of tensile stress at the root of any weld.

9.12 Prohibited Types of Joints and Welds

9.12.1 The joints and welds listed in the following paragraphs are prohibited.

9.12.1.1 Butt joints not fully welded throughout their cross section.

9.12.1.2 Groove welds made from one side only.

(1) Without any backing

(2) With backing, other than steel, that has not been qualified in accordance with 5.2

These prohibitions for groove welds made from one side only shall not apply to

(3) Secondary or nonstress carrying members and shoes or other nonstressed appurtenances

(4) Corner joints parallel to the direction of computed stress, between components for built-up members designed primarily for axial stress

9.12.1.3 Intermittent groove welds.

9.12.1.4 Intermittent fillet welds, except as provided in 9.21.3.1.

9.12.1.5 Bevel-grooves and J-grooves in butt joints for other than the horizontal position (see Fig. 2.9.1).

9.12.1.6 Plug and slot welds on primary tension members.

9.13 Combinations of Welds

If two or more of the general types of welds (groove, fillet, plug, slot) are combined in a single joint, their allowable capacity shall be computed with reference to the axis of the group in order to determine the allowable capacity of the combination. See Appendix A. However, such methods of adding individual capacities of welds does not apply to fillet welds reinforcing groove welds.

9.14 Welds in Combination with Rivets and Bolts

In new work, rivets or bolts in combination with welds shall not be considered as sharing the stress, and the welds shall be provided to carry the entire stress for which the connection is designed. Bolts or rivets used in assembly may be left in place if their removal is not specified. If bolts are to be removed, the plans should indicate whether or not holes should be filled and in what manner.

9.15 Details of Fillet Welds

Fillet welds which support a tensile force that is not parallel to the axis of the weld shall not terminate at corners of parts or members, except as allowed by 9.21.5.2(2), but shall be returned continuously, full size around the corner for a length equal to twice the weld size where such return can be made in the same plane. Boxing shall be indicated on design and detail drawings.

9.16 Eccentricity of Connections

9.16.1 Eccentricity between intersecting parts and members shall be avoided insofar as practicable.

9.16.2 In designing welded joints, adequate provision shall be made for bending stresses due to eccentricity, if any, in the disposition and section of base metal parts and in the location and types of welded joints.

9.16.3 For members having symmetrical cross sections, the connection welds shall be arranged symmetrically about the axis of the member, or proper allowance shall be made for unsymmetrical distribution of stresses.

9.16.4 For axially stressed angle members, the center of gravity of the connecting welds shall lie between the line of the center of gravity of the angle's cross section and the center line of the connected leg. If the center of gravity of the connecting weld lies outside of this zone, the total stresses, including those due to the eccentricity from the center of gravity of the angle, shall not exceed those permitted by this code.

9.17 Connections or Splices—Tension and Compression Members

Connections or splices of tension or compression members made by groove welds shall have complete joint penetration welds. Connections or splices made with fillet or plug welds, except as noted in 9.18, shall be designed for an average of the calculated stress and the strength of the member, but not less than 75% of the strength of the member; or if there is repeated application of load, the maximum stress or stress range in such connection or splice shall not exceed the fatigue stress permitted by the applicable general specification.

9.18 Connections or Splices in Compression Members with Milled Joints

If members subject to compression only are spliced with full-milled bearing, the splice material and its welding shall be arranged, unless otherwise stipulated by the applicable general specifications, to hold all parts in alignment and shall be proportioned to carry 50% of the computed stress in the member. Where such members are in full-milled bearing on base plates, there shall be sufficient welding to hold all parts securely in place.

9.19 Connections of Components of Built-up Members

When a member is made of two or more pieces, the pieces shall be connected along their longitudinal joints by sufficient continuous welds to make the pieces act in unison.

9.20 Transition of Thicknesses or Widths at Butt Joints

9.20.1 Butt joints between parts having unequal thicknesses and subject to tensile stress shall have a smooth transition between the offset surfaces at a slope of no more than 1 in 2-1/2 with the surface of either part. The transition may be accomplished by sloping weld surfaces, by chamfering the thicker part, or by a combination of the two methods (see Fig. 9.20.1).

9.20.2 In butt joints between parts of unequal thickness that are subject only to shear or compressive stress, transition of thickness shall be accomplished as specified in 9.20.1 when offset between surfaces at either side of the joint is greater than the thickness of the thinner part

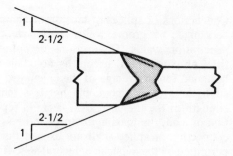
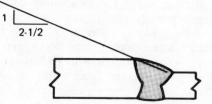

Transition by sloping weld surface

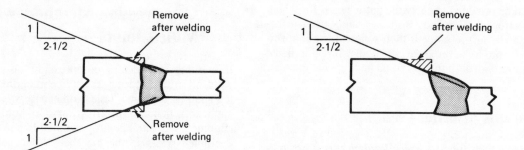

Transition by sloping weld surface and chamfering

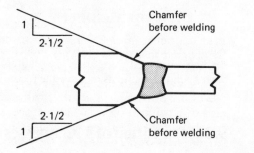
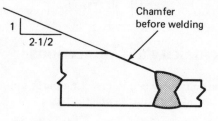

Transition by chamfering thicker part

Center line alignment
(particularly applicable to web plates)

Offset alignment
(particularly applicable to flange plates)

Notes:

1. Groove may be of any permitted or qualified type and detail.

2. Transition slopes shown are the maximum permitted.

Fig. 9.20.1—Transition of thickness at butt joints of parts having unequal thickness

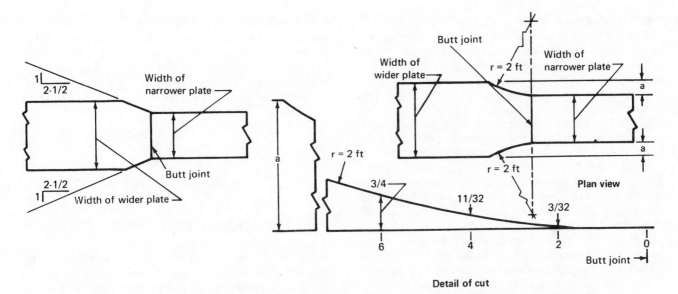

Fig. 9.20.3—Transition of width at butt joints of parts having unequal width

connected. When the offset is equal to or less than the thickness of the thinner part connected, the face of the weld shall be sloped no more than 1 in 2-1/2 from the surface of the thinner part or shall be sloped to the surface of the thicker part if this requires a lesser slope with the following exception: truss member joints and beam and girder flange joints shall be made with smooth transitions of the type specified in 9.20.1.

9.20.3 Butt joints between parts having unequal width and subject to tensile stress shall have a smooth transition between offset edges at a slope of no more than 1 in 2-1/2 with the edge of either part or shall be transitioned with a 2.0 ft (610 mm) radius tangent to the narrower part of the center of the butt joints (see Fig. 9.20.3).

9.21 Girders and Beams

9.21.1 Connections or splices in beams or girders when made by groove welds shall have complete joint penetration welds. Connections or splices made with fillet or plug welds shall be designed for the average of the calculated stress and the strength of the member, but no less than 75% of the strength of member, or if there is repeated application of load, the maximum stress or stress range in such connection or splice shall not exceed the fatigue stress permitted by the applicable general specification.

9.21.2 Splices between sections of rolled beams or built-up girders shall preferably be made in a single transverse plane. Shop splices of webs and flanges in built-up girders, made before the webs and flanges are joined to

each other, may be located in a single transverse plane or multiple transverse planes, but the fatigue stress provisions of the general specification shall apply.

9.21.3 Stiffeners

9.21.3.1 Intermittent fillet welds used to connect stiffeners to beams and girders shall comply with the following requirements:

(1) Minimum length of each weld shall be 1-1/2 in. (38.1 mm).

(2) Welds shall be made on both sides of the joint for at least 25% of its length.

(3) Maximum end-to-end clear spacing of welds shall be twelve times the thickness of the thinner part but not more than 6 in. (152 mm).

(4) Each end of the stiffener, connected to the web, shall be welded on both sides of the joint.

9.21.3.2 Stiffeners, if used, shall preferably be arranged in pairs on opposite sides of the web. Stiffeners may be welded to tension or compression flanges. The fatigue stress or stress ranges at the points of attachment to the tension flange or tension portions of the web shall comply with the fatigue requirements of the general specification. Transverse fillet welds may be used for welding stiffeners to flanges.

9.21.3.3 If stiffeners are used on only one side of the web, they shall be welded to the compression flange.

9.21.4 Girders (built-up I sections) shall preferably be made with one plate in each flange, i.e., without cover plates. The unsupported projection of a flange shall be no more than permitted by the applicable general specification. The thickness and width of a flange may be varied by butt welding parts of different thickness or width with transitions conforming to the requirements of 9.20.

9.21.5 Cover Plates

9.21.5.1 Cover plates shall preferably be limited to one on any flange. The maximum thickness of cover plates on a flange (total thickness of all cover plates if more than one is used) shall be no greater than 1-1/2 times the thickness of the flange to which the cover plate is attached. The thickness and width of a cover plate may be varied by butt welding parts of different thickness or width with transitions conforming to the requirements of 9.20. Such plates shall be assembled and welds ground smooth before being attached to the flange. The width of a cover plate, with recognition of dimensional tolerances allowed by ASTM A6, shall allow suitable space for a fillet weld along each edge of the joint between the flange and the plate cover.

9.21.5.2 Any partial length cover plate shall extend beyond the theoretical end[34] by the terminal distance, or it shall extend to a section where the stress or stress range in the beam flange is equal to the allowable fatigue stress permitted by the applicable general specification, whichever is greater. The theoretical end of the cover plate is the section at which the stress in the flange without that cover plate equals the allowable stress exclusive of fatigue considerations. The terminal distance beyond the theoretical end shall be at least sufficient to allow terminal development in one of the following manners:

(1) Preferably, terminal development shall be made with the end of the cover plate cut square, with no reduction of width in the terminal development length and with a continuous fillet weld across the end and along both edges of the cover plate or flange to connect the cover plate to the flange. For this condition, the terminal development length, measured from the actual end of the cover plate, shall be 1-1/2 times the width of the cover plate at its theoretical end.

(2) Alternatively, terminal development may be made with no weld across the end of the cover plate provided that all of the following conditions are met:

(a) The terminal development length, measured from the actual end of the cover plate, is twice the width of the cover plate at its theoretical end.

(b) The width of the cover plate is symmetrically tapered to a width no greater than 1/3 the width at the theoretical end, but no less than 3 in. (76.2 mm).

(c) There is a continuous fillet weld along both edges of the plate in the tapered terminal development length to connect it to the flange.

9.21.5.3 Fillet welds connecting a cover plate to the flange in the region between terminal developments shall be continuous welds of sufficient size to transmit the incremental longitudinal shear[35] between the cover plate and the flange. Fillet welds in each terminal development shall be of sufficient size to develop the cover plate's portion of the stress in the beam or girder at the inner end of the terminal development length[36] and in no case shall the welds be smaller than the minimum size permitted by 2.7.1.1.

Part D
Workmanship

9.22 Preparation of Material

9.22.1 Edges of material thicker than specified in the following list shall be trimmed if and as required to produce a satisfactory welding edge wherever a weld along the edge is to carry calculated stress:

Sheared edges of material
thicker than   1/2 in. (12.7 mm)

Rolled edges of plates
(other than universal mill plates)
thicker than   3/8 in. (9.5 mm)

Toes of angles or rolled shapes
(other than wide flange sections)
thicker than   5/8 in. (15.9 mm)

Universal mill plates or edges of
flanges of wide flange sections
thicker than   1in. (25.4 mm)

---

34. Defined in AASHTO Standard Specifications for Highway Bridges. (It is also called the "theoretical cut-off point.")

35. The incremental longitudinal shear is VQ/I, where
   V is the vertical shear at the point of calculation.
   Q is the static moment of the cover plate at the point of calculation, taken about the neutral axis of the cover-plated section.
   I is the moment of intertia of the cover-plated section at the point of calculation.

36. This portion of the stress is equal to MQ/I, where
   M is the bending moment at the inner end of the terminal development length.
   Q is the static moment of the cover plate at the inner end of the terminal development length, taken about the neutral axis of the cover-plated section.
   I is the moment of inertia of the cover-plated section at the inner end of the terminal development length.

Commonly, the inner end of the terminal development length will be at the theoretical end of the cover plate but, in the case of a cover plate extension beyond the theoretical end which is greater than the terminal development length, only the length specified in 9.21.5.2(1) or 9.21.5.2(2), whichever is applicable, may be considered in calculating the size of the terminal development welds. Failure to recognize this limitation can result in welds that are too small to support the flange-to-cover plate terminal transition stresses.

The form of edge preparation for butt joints shall conform to the requirements of 2.9 except as modified by 2.6.2.

9.22.2 Steel and weld metal may be oxygen-cut, provided a smooth and regular surface free from cracks and notches is secured, and provided that an accurate profile is secured by the use of a mechanical guide. Freehand oxygen cutting shall be done only where approved by the Engineer.

9.23 Dimensional Tolerances

The dimensions of structural members shall be within the tolerances specified in 3.5. In addition:

9.23.1 Variations from the flatness of girder webs shall be determined by measuring offsets from a straight edge whose length is no less than the least dimension of any panel. The straight edge shall be placed in any position of maximum variation on the web with the ends of the straight edge adjacent to opposite panel boundaries.

9.23.2 Variation from flatness of webs having a depth, D, and a thickness, t, in panels bounded by stiffeners or flanges, or both, whose least panel dimension is d shall not exceed the following:
Intermediate stiffeners on both sides of web
    Interior girders—
      where $D/t < 150$—maximum variation $= d/115$
      where $D/t > 150$—maximum variation $= d/92$
    Fascia girders—
      where $D/t < 150$—maximum variation $= d/130$
      where $D/t > 150$—maximum variation $= d/105$
Intermediate stiffeners on one side only of web
    Interior girders—
      where $D/t < 100$—maximum variation $= d/100$
      where $D/t > 100$—maximum variation $= d/67$
    Fascia girders—
      where $D/t < 100$—maximum variation $= d/120$
      where $D/t > 100$—maximum variation $= d/80$
No intermediate stiffeners—maximum variation $= D/150$

(See Appendix H for tabulation.)

9.23.3 Web distortions of twice the allowable tolerances of 9.23.2 shall be satisfactory when occurring at the end of a girder which has been drilled, or subpunched and reamed; either during assembly or to a template for a field bolted splice; provided, when the splice plates are bolted, the web assumes the proper dimensional tolerances.

9.23.4 If architectural considerations require tolerances more restrictive than described above, specific reference must be included in the bid documents.

9.24 Temporary Welds

Temporary welds shall be subject to the same welding procedure requirements as the final welds. They shall be removed unless otherwise permitted by the Engineer. When they are removed, the surface shall be made flush with the original surface. There shall be no temporary welds in tension zones of members made of quenched and tempered steel except at locations more than 1/6 of the depth of the web from tension flanges of beams or girders. Temporary welds at other locations shall be shown on shop drawings and shall be made with E70XX low-hydrogen electrodes.

9.25 Quality of Welds

9.25.1 Visual Inspection. All welds shall be visually inspected. A weld shall be acceptable by visual inspection if it shows that
    9.25.1.1 The weld has no cracks.
    9.25.1.2 Thorough fusion exists between adjacent layers of weld metal and between weld metal and base metal.
    9.25.1.3 All craters are filled to the full cross section of the weld.
    9.25.1.4 Weld profiles are in accordance with 3.6.
    9.25.1.5 The frequency of piping porosity in fillet welds shall not exceed one in each 4 in. (102 mm) of length and the maximum diameter shall not exceed 3/32 in. (2.4 mm). Exception: for fillet welds connecting stiffeners to web, the sum of the diameters of piping porosity shall not exceed 3/8 in. (9.5 mm) in any linear inch of weld and shall not exceed 3/4 in. (19.0 mm) in any 12 in. (305 mm) length of weld.
    9.25.1.6 A fillet weld in any single continuous weld shall be permitted to underrun the nominal fillet weld size required by 1-/16 in. (1.6 mm) without correction, provided that the undersize portion of the weld does not exceed 10% of the length of the weld. On web-to-flange welds on girders, no underrun is permitted at the ends for a length equal to twice the width of the flange.
    9.25.1.7 Complete joint penetration groove welds in butt joints transverse to the direction of computed tensile stress shall have no piping porosity. For all other groove welds, the frequency of piping porosity shall not exceed one in 4 in. (102 mm) of length and the maximum diameter shall not exceed 3/32 in. (2.4 mm).
    9.25.1.8 Visual inspection of welds in all steels may begin immediately after the completed welds have cooled to ambient temperature. Acceptance criteria for ASTM A514 and A517 steels shall be based on visual inspection performed not less than 48 hours after completion of the weld.

    9.25.2 Radiographic and Magnetic Particle Inspection. Welds that are subject to radiographic or magnetic

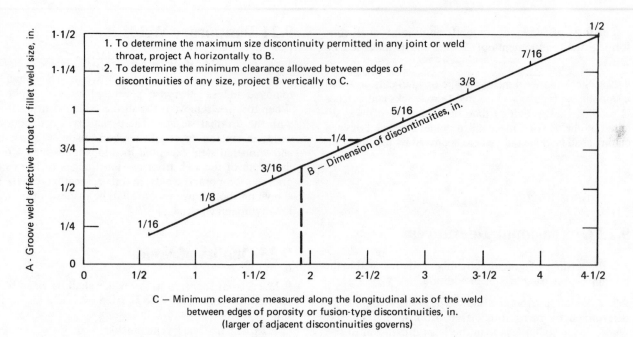

1. To determine the maximum size discontinuity permitted in any joint or weld throat, project A horizontally to B.
2. To determine the minimum clearance allowed between edges of discontinuities of any size, project B vertically to C.

A - Groove weld effective throat or fillet weld size, in.

B – Dimension of discontinuities, in.

C – Minimum clearance measured along the longitudinal axis of the weld between edges of porosity or fusion-type discontinuities, in.
(larger of adjacent discontinuities governs)

Note: Adjacent discontinuities, spaced less than the minimum spacing required by Fig. 9.25.2.1, shall be measured as one length equal to the sum of the total length of the discontinuities plus the length of the space between them and evaluated as a single discontinuity.

Fig. 9.25.2.1—Weld quality requirements for discontinuities occurring in tension welds (limitation of porosity and fusion-type discontinuities)

particle testing in addition to visual inspection shall have no cracks and shall be unacceptable if the radiograph or magnetic particle testing shows any of the types of discontinuities given in 9.25.2.1, 9.25.2.2, 9.25.2.3, or 9.25.2.4.

9.25.2.1 For welds subject to tensile stress under any condition of loading, the greatest dimension of any porosity or fusion-type discontinuity that is 1/16 in. (1.6 mm) or larger in greatest dimension shall not exceed the size, B, indicated in Fig. 9.25.2.1 for the effective throat or weld size involved. The distance from any porosity or fusion-type discontinuity described above to another such discontinuity, to an edge, or to any intersecting weld shall not be less than the minimum clearance allowed, C, indicated by Fig. 9.25.2.1 for the size of discontinuity under examination.

9.25.2.2 For welds subject to compressive stress only and specifically indicated as such on the design drawings, the greatest dimension of porosity or a fusion-type discontinuity that is 1/8 in. (3.2 mm) or larger in greatest dimension shall not exceed the size, B, nor shall the space between adjacent discontinuities be less than the minimum clearance allowed, C, indicated by Fig. 9.25.2.2 for the size of discontinuity under examination.

9.25.2.3 Independent of the requirements of 9.25.2.1 and 9.25.2.2, discontinuities having a greatest dimension

of less than 1/16 in. (1.6 mm) shall be unacceptable if the sum of their greatest dimension exceeds 3/8 in. (9.5 mm) in any linear inch of weld.

9.25.2.4 The limitations given by Figs. 9.25.2.1 and 9.25.2.2 for 1-1/2 in. (38.1 mm) joint effective throat shall apply to all joints or effective throats of greater thickness.

9.25.2.5 Appendix F illustrates the application of the requirements given in 9.25.2.1.

9.25.3 Ultrasonic Inspection

9.25.3.1 Welds that are subject to ultrasonic testing in addition to visual inspection are acceptable if they meet the following requirements:

(1) Welds subject to tensile stress under any condition of loading shall conform to the requirements of Table 9.25.3.

(2) Welds subject to compressive stress shall conform to the requirements of Table 8.15.3.

Ultrasonically tested welds are evaluated on the basis of a discontinuity reflecting ultrasound in proportion to its effect on the integrity of the weld.

9.25.3.2 Complete joint penetration welds joining web-to-flange in plate girders shall conform to the requirements of Table 8.15.3. When such welds are subject to calculated tensile stress normal to the weld throat they

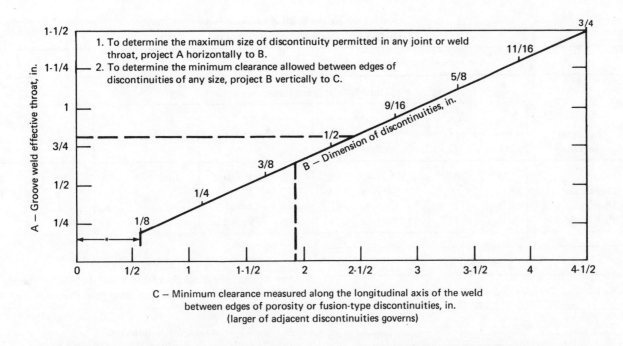

C — Minimum clearance measured along the longitudinal axis of the weld
between edges of porosity or fusion-type discontinuities, in.
(larger of adjacent discontinuities governs)

*The maximum size of a discontinuity located within this distance from an edge of plate shall be 1/8 in. (3.2 mm), but a 1/8 in. discontinuity must be 1/4 in. (6.4 mm) or more away from the edge. The sum of discontinuities less than 1/8 in. in size and located within this distance from the edge shall not exceed 3/16 in. (4.8 mm). Discontinuities 1/16 in. to less than 1/8 in. will not be restricted in other locations unless they are separated by less than 2 L (L being the length of the larger discontinuity); in which case, the discontinuities shall be measured as one length equal to the total length of the discontinuities and space and evaluated as shown in Fig. 9.25.2.2.

Fig. 9.25.2.2—Weld quality requirements for discontinuities occurring in compression welds (limitations of porosity or fusion-type discontinuities)

should be designated as 'tension welds' on the design drawings. Complete joint penetration web-to-flange welds designated as tension welds on design drawings shall conform to the requirements of Table 9.25.3.

9.25.4 Liquid Penetrant Inspection. Welds that are subject to liquid penetrant testing, in addition to visual inspection, shall be evaluated on the basis of the requirements for visual inspection.

9.25.5 When welds are subject to nondestructive testing, in accordance with 9.25.2, 9.25.3, and 9.25.4, the testing may begin immediately after the completed welds have cooled to ambient temperature. Acceptance criteria for ASTM A514 and A517 steels shall be based on nondestructive testing performed not less than 48 hours after completion of the welds.

Table 9.25.3
Ultrasonic acceptance criteria

Minimum acceptance levels (decibels)

Reflector severity	Weld thickness (in.) and transducer angle													
	5/16 to 3/4	>3/4 to 1-1/2	>1-1/2 to 2-1/2			>2-1/2 to 4			>4 to 6			>6 to 8		
	70°	70°	70°	60°	45°	70°	60°	45°	70°	60°	45°	70°	60°	45°
Large reflectors	+14	+ 9	+ 5	+ 8	+10	+ 2	+ 5	+ 7	− 1	+ 2	+ 4	− 3	0	+ 2
Small reflectors	+15	+10	+ 7	+10	+12	+ 4	+ 7	+ 9	+ 1	+ 4	+ 6	− 1	+ 2	+ 4
Minor reflectors	+16	+11	+ 9	+12	+14	+ 6	+ 9	+11	+ 3	+ 6	+ 8	+ 1	+ 4	+ 6

Scanning levels

Sound path distance

in.	mm	Above zero reference, dB
To 2-1/2	63.5	+20
>2-1/2 to 5	63.5 - 127	+25
>5 to 10	127 - 254	+35
>10 to 15	254 - 381	+45

Large reflectors

Any discontinuity, REGARDLESS OF LENGTH, having a more serious rating (smaller number) than this level shall be rejected.

Small reflectors

Any discontinuity longer than 3/4 in. (19 mm) having a more serious rating (smaller number) than this level shall be rejected.

Minor reflectors

Any discontinuity longer than 2 in. (51 mm) having a more serious rating (smaller number) than this level shall be rejected.

Notes:

1. Discontinuities which have a more serious rating than those of 'minor reflectors' shall be separated by at least 2L, L being the length of the larger discontinuity. Discontinuities not separated by at least 2L are considered to be one continuous discontinuity whose length is determined by the combined length of the discontinuities plus their separation distance.
2. Discontinuities which have a more serious rating than those of 'minor reflectors' shall not begin at a distance smaller than 2L from the end of the weld, L being the discontinuity length.
3. Discontinuities detected at 'scanning levels' in the root-face areas of complete joint penetration double-V-groove welds, double-J-groove welds, double-U-groove welds, and double-bevel-groove welds shall be evaluated at an acceptance level of 4 dB* more sensitive than prescribed in this table when such welds are designated on design drawings as 'tension welds.'
4. Discontinuities which have a more serious rating than those of 'minor reflectors' and which have a length greater than 3/4 in. (19 mm) and less than 2 in. (51 mm) are permitted in the middle half of the weld thickness.

*i.e., add +4 dB to the number in the table.

10. Design of New Tubular[37] Structures

Part A
General Requirements

10.1 Application

10.1.1 This section supplements Sections 1 through 6 and is to be used in conjunction with the prescribed specification for the design and construction of steel structures in which the loads are carried primarily by tubular members. It is not intended to apply to pressure vessels or pressure piping.

10.1.2 Members in tubular structures shall be identified as shown in Fig. 10.1.2.

10.1.3 Symbols used in Section 10 are as follows:

Symbol	Meaning	Reference
a	major width of rectangular hollow section product	Footnote 37
a_x	ratio of a to $\sin \theta$	10.8.5
b	minor width of rectangular tubes	Footnote 37

---

37. "Tubular" products is a generic term for a family of hollow sections of various cross-sectional configurations. The products dealt with in this section are made of steels listed in 10.2. The term pipe as used in Section 5 denotes cylindrical products to differentiate from square and rectangular hollow section products. However, a tube or tubing can also be cylindrical. Users should note the AISC designations of tubular sections. e.g.,

TS$D \times t$		for circular tubes
TS$a \times b \times t$		for square and rectangular tubes (referred to collectively as box sections herein)

where,		
	TS	is the group symbol
	t	is the wall thickness
	D	is the OD
	a	is the major width
	b	is the minor width

Symbol	Meaning	Reference
β	(beta) diameter ratio of D_b to D_m	10.13.1.5; Fig. 10.13.1.3
	ratio of r_b to R (circular sections)	10.8.4
	ratio of b to D; (box sections)	Fig. 10.1.2 (M)
C	outside dimension of rectangular member	Fig. 10.1.2 (M)
D	outside diameter OD (circular tubes) or	Footnote 37
	outside width of main member (box sections)	Fig. 10.1.2 (M)
D	cumulative fatigue damage ratio, $\sum \frac{n}{N}$	10.7.4.2
D_b	diameter of branch member	Fig. 10.13.1.3
D_m	diameter of main member	Fig. 10.13.1.3
η	(eta) ratio of a_x to D	Fig. 10.1.2 (B); Fig. 10.1.2 (M) (Table)
ϵ_{TR}	(epsilon) total strain range	Fig. 10.7.4
F_1	web crippling capacity of main member (matched connections); punching shear at material limit (stepped connections)	10.5.1.1 (1)
F_2	punching shear capacity of main member	10.5.1.1 (2)
F_y	yield strength of base metal	Table 10.5.1; Footnote 38
f_a	axial stress in branch member	Fig. 10.5.1 (A)
f_b	bending stress in branch member	Footnote 39; Fig. 10.5.1 (A)
f_{by}	nominal stress, in-plane bending	Footnote 39
f_{bz}	nominal stress, out-of-plane bending	Footnote 39
γ	(gamma) main member flexibility parameter ratio R to t_c (circular sections);	Fig.10.1.2 (M) (Table)
	ratio of D to $2t_c$ (box sections)	Fig. 10.1.2 (M) Table

Symbol	Meaning	Reference
γ_b	radius thickness ratio of tube at transition	Table 10.7.3 (Note 4)
ID	inside diameter	Fig. 10.12 B (F)
K	connection configuration	Fig. 10.1.2 (E)
K_a	relative length factor	10.5.1; Fig. 10.5.1B; 10.8.4
K_b	relative section factor	10.5.1; Fig. 10.5.1B; 10.8.4
L	size of fillet	Fig. 10.11.3; Fig. 10.13.1.1(A) Detail A; Fig. 10.13.1.3
	length of joint can	10.5.2.1(2)
l_1	actual weld length where branch contacts main member	10.5.1.5 (1)
l_2	projected chord length (one side) of overlapping weld	Fig. 10.5.1.4
N	cycles of load	Fig. 10.7.4
n	number of cycles at given stress range	10.7.4.2
OD	outside diameter	Footnote 37; Fig. 10.12 B (E)
ω	(omega) end preparation angle	Fig. 10.13.1.1(A) Details A,B,C
P	axial load component	Fig. 10.5.1.4
	compressive branch member axial load	10.5.2.1(1)
$P_\perp$	individual member load component perpendicular to main member axis	10.5.1.5(1); Fig. 10.5.1.4
Φ	(phi) joint included angle	Fig. 10.1.2(D) (E); Fig. 10.13.1.1A Details
π	(pi) ratio of circumference to diameter of circle	10.8.4
Ψ	(psi) dihedral angle	Fig. 10.1.2(K) (M) Fig. 10.13.1.1 A
$\overline{\Psi}$	(psi bar) supplementary angle to the local dihedral angle	Fig. 10.1.2(K); 10.5.1.6 (Note)
	angle change at transition	Table 10.7.3 (Note 4)
Q	unbalanced radial line load	10.5.1.6
$Q\beta$	geometry modifier	Table 10.5.1
Q_f	stress interaction term	Table 10.5.1
R	outside radius, main member	10.8.4
R	root opening (joint fit-up)	Fig. 10.13.1.1A
r	effective radius of intersection	10.8.4
r_b	radius of branch	Fig. 10.5.1 A
SCF	stress concentration factor	Table 10.7.3 (Note 4)
$\sum l_1$	(sigma) summation of actual weld lengths	10.5.1.5(2)
T	connection configuration	Fig. 10.1.2(C)

Symbol	Meaning	Reference
TCBR	total range of nominal axial and bending stress	Table 10.7.3
TS	group symbol for tubular sections	Footnote 37
	tension stress	Table 10.7.3
	throat of weld	Fig. 10.10.1(A)
t	tubular member wall thickness: main member for punching shear calculations	Fig. 10.5.1.5(1)
	main member	10.5.2.1(1)
	branch member for dimensioning of complete joint penetration groove welds	Fig. 10.13.1.1A (details A, B, C, D)
	thinner member for dimensioning partial penetration groove welds and fillet welds	Fig. 10.13.1.2A
	wall thickness	Footnote 37
t_b	wall thickness of thinner branch member	Fig. 10.5.1 A
	wall thickness of branch member	Fig. 10.5.1.4
t_c	wall thickness of main member	Fig. 10.1.2(M)
	joint can thickness	10.5.2.1 (2)
t_w	the lesser of weld effective throat or t_b	10.5.1.5 (1)
τ	(tau) branch-to-main relative thickness geometry parameter; ratio of t_b to t_c	Fig. 10.1.2 (M)
θ	(theta) acute angle between two member axes	10.8.4; Fig. 10.1.2(D), (E); Fig. 10.5.1 A
	angle between member center lines	Fig. 10.1.2(M) (Table)
	brace intersection angle	Fig. 10.5.1 B
U	utilization ratio of (axial and bending stress) to allowable stress, at point under consideration	Table 10.5.1
V_p	punching shear stress	Table 10.5.1; Fig. 10.5.1 A
V_w	allowable shear stress for weld between branch members	10.5.1.5 (1) Fig. 10.5.1.4
W	backup weld throat	Fig. 10.13.1.1A (Details C,D)
x	algebraic variable $\dfrac{1}{2\pi\sin\theta}$	10.8.4
Y	connection configuration	Fig. 10.1.2 (D)
y	algebraic variable $\dfrac{1}{3\pi}\cdot\dfrac{3-\beta^2}{2-\beta^2}$	10.8.4

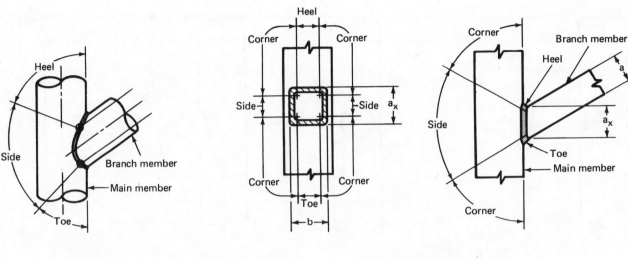

(A) Circular sections (B) Box sections

Fig. 10.1.2—Parts of a tubular connection

10.2 Base Metal

10.2.1 Steel base metal to be used for welded tubular structures shall conform to the requirements of the latest edition of any specification listed below. Combinations of approved steel base metals may be welded together.

10.2.1.1 ASTM A36, Specification for Structural Steel

10.2.1.2 ASTM A53, Grade B, Specification for Pipe, Steel, Black and Hot-Dipped, Zinc-Coated, Welded and Seamless

10.2.1.3 ASTM A106, Grade B, Specification for Seamless Carbon Steel Pipe for High-Temperature Service

10.2.1.4 ASTM A131, Specification for Structural Steel for Ships

10.2.1.5 ASTM A139, Grade B, Specification for Electric-Fusion (Arc)-Welded Steel Pipe (Sizes 4 in. and Over)

10.2.1.6 ASTM A242, Specification for High-Strength Low-Alloy Structural Steel (if the properties are suitable for welding)

10.2.1.7 ASTM A381, Grade Y-35, Specification for Metal-Arc-Welded Steel Pipe for High-Pressure Transmission Systems

10.2.1.8 ASTM A441, Specification for High-Strength Low-Alloy Structural Manganese-Vanadium Steel

10.2.1.9 ASTM A500, Specification for Cold-Formed Welded and Seamless Carbon Steel Structural Tubing in Rounds and Shapes

10.2.1.10 ASTM A501, Specification for Hot-Formed Welded and Seamless Carbon Steel Structural Tubing

10.2.1.11 ASTM A514, Specification for High-Yield-Strength, Quenched and Tempered Alloy Steel Plate, Suitable for Welding

10.2.1.12 ASTM A516, Specification for Pressure Vessel Plates, Carbon Steel, for Moderate- and Lower-Temperature Service

10.2.1.13 ASTM A517, Specification for Pressure Vessel Plates, Alloy Steel, High-Strength, Quenched and Tempered

10.2.1.14 ASTM A524, Specification for Seamless Carbon Steel Pipe for Process Piping

10.2.1.15 ASTM A529, Specification for Structural Steel with 42 000 psi Minimum Yield Point (1/2 in. Maximum Thickness)

10.2.1.16 ASTM A537, Specification for Pressure Vessel Plates, Carbon-Manganese-Silicon Steel, Heat-Treated

10.2.1.17 ASTM A570, Specification for Hot-Rolled Carbon Steel Sheet and Strip, Structural Quality

10.2.1.18 ASTM A572, Specification for High-Strength Low-Alloy Columbium-Vanadium Steels of Structural Quality

10.2.1.19 ASTM A573, Grade 65, Specification for Structural Carbon Steel Plates of Improved Toughness

10.2.1.20 ASTM A588, Specification for High-Strength Low-Alloy Structural Steel with 50 000 psi Minimum Yield Point to 4 in. Thick

10.2.1.21 ASTM A595, Specification for Steel Tubes, Low-Carbon, Tapered for Structural Use

10.2.1.22 ASTM A606, Type 2 (Type 4 if the properties are suitable for welding), Specification for Steel Sheet and Strip, Hot-Rolled and Cold-Rolled, High-Strength, Low-Alloy, with Improved Corrosion Resistance

10.2.1.23 ASTM A607, Grades 45, 50, and 55, Specification for Steel Sheet and Strip, Hot-Rolled and Cold-Rolled, High-Strength, Low-Alloy Columbium and/or Vanadium

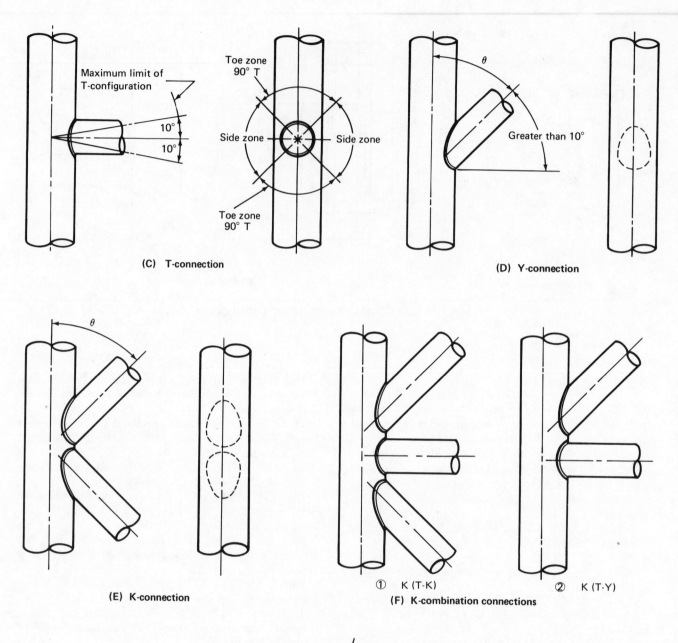

(C) T-connection

(D) Y-connection

(E) K-connection

① K (T-K)

② K (T-Y)

(F) K-combination connections

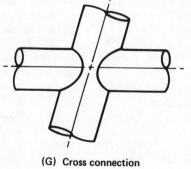

(G) Cross connection

Fig. 10.1.2 (continued)—Parts of a tubular connection

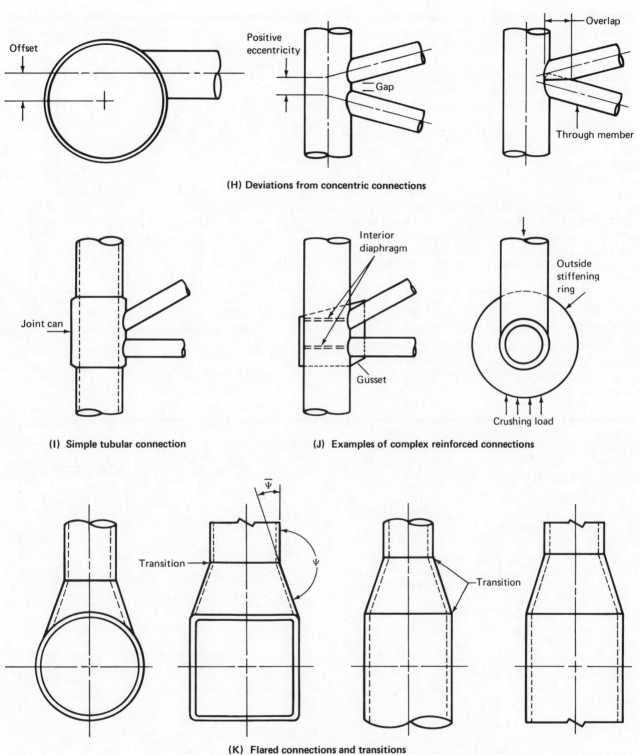

(H) Deviations from concentric connections

(I) Simple tubular connection

(J) Examples of complex reinforced connections

(K) Flared connections and transitions

Fig. 10.1.2 (continued) — Parts of a tubular connection

10.2.1.24 ASTM A618, Grades II and III (Grade I if the properties are suitable for welding), Specification for Hot-Formed Welded and Seamless High-Strength Low-Alloy Structural Tubing

10.2.1.25 ASTM A633, Specification for Normalized High-Strength Low-Alloy Structural Steel

10.2.1.26 ASTM A709, Specification for Structural Steel for Bridges

10.2.1.27 API 5L, Grade B, Specification for Line Pipe

10.2.1.28 API 5LX, Grade X42, Specification for High-Test Line Pipe

10.2.1.29 API 2B (when made from plate steel listed herein), Specification for Fabricated Structural Steel Pipe

10.2.1.30 API 2H, Specification for Carbon-Manganese Steel Plate for Offshore Platform Tubular Joints

10.2.1.31 ABS Grades A, B, D, E, DS, and CS, Requirements for Ordinary-Strength Hull Structural Steel

10.2.1.32 ABS Grades AH32, DH32, EH32, AH36, DH36, and EH36, Requirements for Higher-Strength Hull Structural Steel

10.2.2 When an ASTM A709 grade of structural steel is considered for use, its weldability shall be established by the steel producer, and the procedure for welding it shall be established by qualification in accordance with the requirements of 5.2 and other such requirements as prescribed by the Engineer with the following exception: if the grade to be supplied will meet the chemical and mechanical properties of ASTM A36, A572 Grade 50, A588, or A514, the applicable prequalified procedures of this code shall apply.

10.2.3 Base Metals Not Approved in 10.2.1

10.2.3.1 When a steel other than those listed in 10.2.1 is approved under the provisions of the general specifications and is proposed for welded construction, the weldability of the steel and the joint welding procedure shall be established by qualification in accordance with the requirements of 5.2 and such other requirements as prescribed by the Engineer.

10.2.3.2 The responsibility for determining weldability, including the assumption of additional testing costs, is assigned to the party who either specifies a material not listed in 10.2.1 or who proposes the use of a substitute material not listed in 10.2.1. The fabricator shall have the responsibility for establishing the joint welding procedure by qualification.

10.2.4 Extension bars and runoff plates used in welding shall conform to the following requirements:

(1) When used in welding with an approved steel listed in 10.2.1, they may be any of the steels listed in 10.2.1

(2) When used in welding with a steel qualified in accordance with 10.2.3, they may be

(a) The steel qualified, or
(b) Any steel listed in 10.2.1

Steel for backing shall conform to the requirements of (1) and (2), except that 100 ksi minimum yield strength steel as backing shall only be used with 100 ksi minimum yield strength steel.

Spacers shall be of the same material as the base metal.

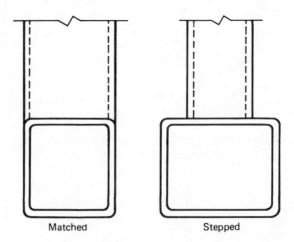

(L) Connection types for box sections

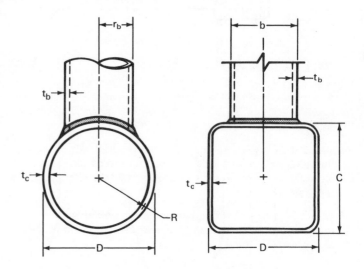

(M) Geometric parameters

Parameter	Circular sections	Box sections
β	r_b/R	b/D
η	—	a_x/D See Fig. 10.1.2 (B)
γ	R/t_c	$D/2t_c$
τ	t_b/t_c	t_b/t_c
θ	Angle between member center lines	
Ψ	Local dihedral angle at given point on welded joint	

Fig. 10.1.2 (continued)—Parts of a tubular connection

10.2.5 The provisions of this code are not intended for use with steels having a specified minimum yield point or yield strength over 100 000 psi (690 MPa).

Part B
Allowable Unit Stresses

10.3 Base Metal Stresses

The base metal stresses shall be those specified in the applicable design specifications, with the following modifications:

10.3.1 For circular sections having D/t greater than $3300/F_y$,[38] the possibility of local buckling at axial compressive stresses less than the yield strength shall be considered.

10.3.2 Moments caused by significant deviation from concentric connections shall be provided for in analysis and design. See Fig. 10.1.2(H).

10.4 Unit Stresses in Welds

10.4.1 Except as modified in 10.5, 10.6, and 10.7, the allowable stresses in welds shall be as shown in Table 10.4.1.

10.4.2 Fiber stresses due to bending shall not exceed the values prescribed for tension and compression, unless the weld is proportioned to develop fully the strength of sections joined.

10.4.3 Plug or slot welds shall not be ascribed any value in resistance to stress other than shear in the plane of the faying surfaces.

10.5 Limitations of the Strength of Welded Tubular Connections

10.5.1 Local Failure. Where a STEPPED BOX or CIRCULAR T-, Y-, or K-connection is made by simply welding the branch member to the main member, local stresses at a potential failure surface through the main member wall may limit the usable strength of the welded joint. The shear stress at which such failure occurs depends not only upon the strength of the main member steel, but also on the geometry of the connection. Such connections shall be proportioned so that the punching

38. F_y is the yield strength (ksi) of the base metal.

shear stress on the potential failure surface (see Fig. 10.5.1) as given by

$$\text{acting } V_p = \tau \left[\frac{f_a \sin \theta}{K_a} + \frac{f_b}{K_b} \right]$$

does not exceed the allowable shear stress specified in the applicable design specification (e.g., $0.40F_y$), nor a reduced allowable V_p, as given by

$$\text{allowable } V_p = Q_\beta Q_f \times (\text{basic } V_p)$$

Terms used in the foregoing equations are defined as follows:

τ, θ, and other parameters of connection geometry are defined in Fig. 10.1.2(M).

f_a, f_b are nominal axial and nominal bending stresses respectively in a branch member.[39]

K_b, K_a are relative length and section factors as given in Fig. 10.5.1(B) and 10.8.3.

(Basic V_p) = basic allowable punching shear as given in Table 10.5.1.

Q_β, Q_f are geometry modifier and stress interaction terms, respectively, also given in Table 10.5.1.

10.5.1.1 For MATCHED CONNECTIONS of BOX SECTIONS as defined in Fig. 10.1.2(L), allowable static capacity for loads normal to the main member shall be taken as the sum of F_1 and F_2 as follows:

(1) Along the sides, web crippling capacity of the main member

$$F_1 = 2t_c a_x (0.60 Q_f F_y)$$

(2) Along the heel and toe, punching shear (taken as single shear) in accordance with 10.5.1, except

$$Q_\beta = 1.25 (1 + \eta) \text{ for } \eta < 1$$

$$Q_\beta = 2.5 \text{ for } \eta \geq 1$$

where $\eta = a_x/D$ (see Fig. 10.1.2B)

thus $F_2 = 2t_c b \left[Q_\beta \cdot Q_f \cdot \frac{F_y}{0.6\gamma} \right]$

Note: Q_f shall be as shown in Table 10.5.1.

10.5.1.2 For MATCHED CONNECTIONS of BOX SECTIONS conforming to all of the requirements given below, the full static load capacity of the branch members shall be assumed:

(1) Complete joint penetration groove weld conforming to Fig. 10.13.1.1(B) and matching weld metal (Table 4.1.1)

39. For bending about two axes (e.g., y and z), the effective resultant bending stress in CIRCULAR and SQUARE BOX SECTIONS may be taken as
$$f_y = \sqrt{f_{by}^2 + f_{bz}^2}$$

Table 10.4.1
Allowable stresses in welds

Type of weld	Tubular application	Kind of stress	Permissible unit stress	Required weld metal strength level[1]
Complete joint penetration groove weld	Longitudinal seam of tubular members	Tension or compression parallel to axis of the weld[2]	Same as for base metal[1]	Weld metal with a strength level equal to or less than matching weld metal may be used.
	Butt splices of tubular members	Compression normal to the effective area[2]	Same as for base metal	Weld metal with a strength level equal to or less than matching weld metal may be used.
		Tension or shear on effective area	Same as for base metal	Matching weld metal must be used. See Table 4.1.1.
	Structural T-,Y-, or K-connections in structures designed for critical loading. such as fatigue, which would normally call for complete joint penetration welds.	Tension, compression, or shear on base metal adjoining weld conforming to detail of Fig. 10.13.1.1 (tubular weld made from outside only).	Same as base metal or as limited by connection geometry (see 10.5)	Matching weld metal must be used. See Table 4.1.1.
		Tension, compression, or shear on effective area of groove welds, made conventionally from both sides or with backing.		Matching weld metal must be used. See Table 4.1.1.

1. For matching weld metal see Table 4.1.1.
2. Beam or torsional shear up to 0.30 minimum specified tensile strength of weld metal is permitted, except that shear on adjoining base metal shall not exceed $0.40F_y$.
3. Groove and fillet welds parallel to the longitudinal axis of tension or compression members, except in connection areas, are not considered as transferring stress and hence may take the same stress as that in the base metal, regardless of electrode (filler metal) classification. Where the provisions of 10.5.1 are applied, seams in the main member within the connection area shall be complete joint penetration groove welds with matching filler metal, as defined in Table 4.1.1.

Table 10.4.1 (continued)
Allowable stresses in welds

Type of weld	Tubular application	Kind of stress		Permissible unit stress	Required weld metal strength level[1]
Partial joint penetration groove weld	Longitudinal seam of tubular members	Tension or compression parallel to axis of the weld[2]		Same as for base metal[3]	Weld metal with a strength level equal to or less than matching weld metal may be used.
	Butt splices of tubular members	Compression normal to the effective throat[2]	Joint not designed to bear	0.50 × specified minimum tensile strength of weld, except that stress on adjoining base metal shall not exceed $0.60F_y$.	Weld metal with a strength level equal to or less than matching weld metal may be used.
			Joint designed to bear	Same as for base metal	
		Tension or shear on effective throat		0.30 × specified minimum tensile strength of weld metal, except that stress on adjoining base metal shall not exceed $0.50F_y$ for tension, nor $0.40F_y$ for shear.	Weld metal with a strength level equal to or less than matching weld metal may be used.
	Structural T-, Y-, or K-connection in ordinary structures	Load transfer across the weld as stress on the effective throat		0.30 × specified minimum tensile strength of weld metal, or as limited by connection geometry (see 10.5), except that stress on adjoining base metal shall not exceed $0.60F_y$ for tension and compression, nor $0.40F_y$ for shear.	Matching weld metal must be used. See Table 4.1.1.
Fillet weld	Longitudinal seam of tubular members	Tension or compression parallel to axis of the weld		Same as for base metal[3]	Weld metal with a strength level equal to or less than matching weld metal may be used.
	Structural T-, Y-, or K-connection in ordinary structures; lap splice of tubular members.	Shear stress on effective throat regardless of direction of loading		0.30 × specified minimum tensile strength of weld metal, or as limited by connection geometry (see 10.5), except that shear stress on adjoining base metal shall not exceed $0.40F_y$.	Weld metal with a strength level equal to or less than matching weld metal may be used.[4]

1. For matching weld metal see Table 4.1.1.
2. Beam or torsional shear up to 0.30 minimum specified tensile strength of weld metal is permitted, except that shear on adjoining base metal shall not exceed $0.40F_y$.
3. Groove and fillet welds parallel to the longitudinal axis of tension or compression members, except in connection areas, are not considered as transferring stress and hence may take the same stress as that in the base metal, regardless of electrode (filler metal) classification. Where the provisions of 10.5.1 are applied, seams in the main member within the connection area shall be complete joint penetration groove welds with matching filler metal, as defined in Table 4.1.1.
4. See 10.5.3.

Table 10.5.1
Terms for finding the allowable punching shear, V_p

	Circular sections*		Stepped box connections**	
Basic V_p	$\dfrac{F_y}{0.9\gamma^{0.7}}$		$\dfrac{F_y}{0.6\gamma}$	
Geometry modifier Q_β	$\dfrac{0.30}{\beta(1-0.833\beta)}$	for $\beta>0.6$	$\dfrac{0.25}{\beta(1-\beta)}$	for $\beta>0.5$***
	1.0	for $\beta\leqslant0.6$	1.0	for $\beta\leqslant0.5$
Stress interaction term Q_f	1.22 − 0.5U for U >0.44			
	1.0 for U ≤0.44			

Notes:
1. U is the utilization ratio (ratio of actual to allowable) for axial and bending stresses in the main member at the point under consideration, e.g., $U=(f_a+f_b)/0.6\,F_y$ with 1/3 increase applicable to denominator.
2. γ, β are geometry parameters defined by Fig. 10.2 M.
3. F_y = the specified minimum yield strength of the main member steel, but not more than 2/3 the tensile strength.

*For circular cross sections in compression, see 10.5.2.1.
**Higher capacities may be used for small-size light gage tubes when justified by tests.
***See 10.5.1.3 for $\beta>0.8$.

(2) For main member, $D/t<22$, and F_y not less than that of branch member
(3) $\tau\leqslant0.72$ (or $\tau\leqslant1.0$ and main member U≤0.44)
(4) $\eta\geqslant1.0$

10.5.1.3 For STEPPED BOX CONNECTIONS having $\beta>0.8$ or $\beta>\eta$, or both, the allowable static capacity for loads normal to the main member shall be computed in accordance with 10.5.1, but shall not exceed the sum of F_1+F_2.

(1) Along the sides, punching shear at the material limit: $F_1=2t_c\,a_x(0.4F_y)$

(2) Along the heel and toe, punching shear as defined in 10.5.1.1(2)

10.5.1.4 For STEPPED BOX CONNECTIONS having $\beta<0.8$, yield line analysis may be used in lieu of the punching shear method of 10.5.1.

10.5.1.5 OVERLAPPING JOINTS, in which part of the load is transferred directly from one branch member to another through their common weld, shall include the following checks:

(1) The allowable individual member load component, $P_\perp$, perpendicular to the main member axis shall be taken as $P_\perp=(V_p t l_1)+(2V_w t_w l_2)$ where V_p is the allowable punching shear as defined in 10.5.1, and

t = the main member thickness

l_1 = actual weld length for that portion of the branch member which contacts the main member

V_w = allowable shear stress for the weld between the branch members (Table 10.4.1)

t_w = the lesser of the weld effective throat or the thickness, t_b, of the thinner branch member

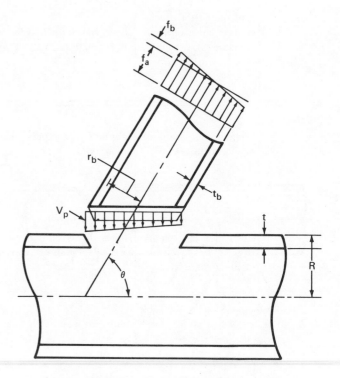

Fig. 10.5.1A—Punching shear stress

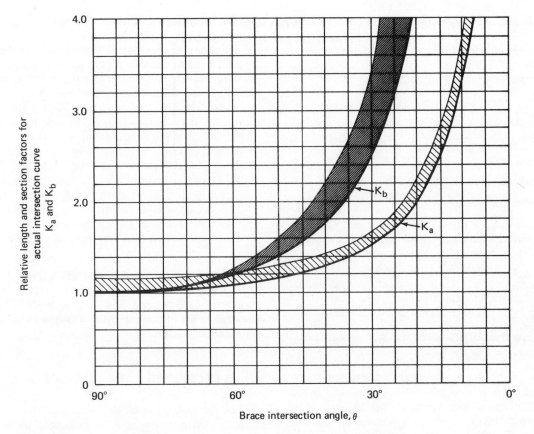

Brace intersection angle, θ

Notes:
1. Solid curves are for circular and square box sections having small β. Shaded areas indicate higher values possible under 10.8.
2. Not prequalified for $\theta < 15°$.

Fig. 10.5.1B — K-factors for calculating punching shear stress

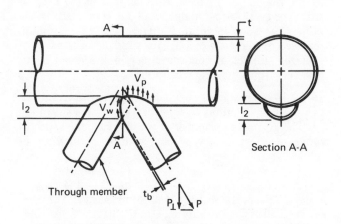

Fig. 10.5.1.5 — Detail of overlapping joint

l_2 = the projected chord length (one side) of the overlapping weld, measured perpendicular to the main member.

These terms are illustrated in Fig. 10.5.1.4.

(2) The allowable combined load component parallel to the main member axis shall not exceed $V_w t_w \sum l_1$, where $\sum l_1$ is the sum of the actual weld lengths for all braces in contact with the main member.

(3) The overlap shall preferably be proportioned for at least 50% of the acting $P_\perp$. In no case shall the branch member wall thickness exceed the main member wall thickness.

(4) Where the branch members carry substantially different loads or one branch member is thicker than the other, or both, the heavier branch member shall preferably be the through member with its full circumference welded to the main member.

10.5.1.6 Flared connections and tube size transitions not excepted below shall be checked for local stresses caused by the change in direction (angle $\overline{\Psi}$) at the transition, using the following assumptions:

Unbalanced radial line load $Q = t(f_a + f_b) \tan \overline{\Psi}$

$$\text{Acting } V_p = \frac{Q}{2t}$$

Allowable $V_p = Q_f \times$ (basic V_p) where t is the thickness of the member and other terms are as defined in 10.5.1.

Exception:

circular tubes having D/t less than 30
box sections having D/t less than 20

And

transition slope is less than 1:2-1/2

Note: $\overline{\Psi}$ is the supplementary angle to the local dihedral angle Ψ; see Fig. 10.1.2(K).

10.5.1.7 For critical connections whose sole failure would be catastrophic, or for architectural applications where localized deformations in the connections would be objectionable, it is recommended that allowable punching shear stresses given in the foregoing sections of 10.5.1 should be reduced by one-third, except where higher loads are justified by tests.

10.5.2 General Collapse. Strength and stability of the main member in a tubular connection acting as a cylindrical shell (and any reinforcement) should be investigated using available technology in accordance with the applicable design code.

10.5.2.1 General collapse is particularly severe in cross connections and connections subjected to crushing loads see Fig. 10.1.2(J). Such connections may be reinforced by increasing the main member thickness, or by the use of diaphragms, rings, or collars.

(1) For unreinforced circular cross connections, the compressive branch member axial load shall not exceed

$$P = \frac{t^2 F_y (1.9 + 7.2\beta) Q_\beta Q_f}{\sin \theta}$$

(2) For circular cross connections reinforced by a "joint can" having increased thickness, t_c, and length, L, the allowable branch axial load, P, may be employed as

$$P = P_{(1)} + \frac{\text{L}}{2.5\text{D}} \left[P_{(2)} - P_{(1)} \right] \text{ for L} < 2.5\text{D}$$

$$P = P_{(2)} \quad \text{for L} \geqslant 2.5\text{D}$$

where $P_{(1)}$ is obtained by using the nominal member thickness, t, in the equation in (1); and, $P_{(2)}$ is obtained by using the joint can thickness, t_c, in the same equation.

10.5.2.2 For K-connections in which the main member thickness required to meet the local shear provisions of 10.5.1 extends at least D/4 beyond the connecting branch member welds, general collapse need not be checked.

10.5.3 Uneven Distribution of Load

10.5.3.1 Due to differences in the relative flexibilities of the main member loaded normal to its surface and the branch member carrying membrane stresses parallel to its surface, transfer of load across the weld is highly non-uniform, and local yielding can be expected before the connection reaches its design load. To prevent progressive failure of the weld and insure ductile behavior of the joint, the minimum welds provided in simple T-, Y-, or K-connections shall be capable of developing, at their ultimate breaking strength, the lesser of the brace member yield strength or local strength (punching shear or web crippling) of the main member.[40]

10.5.3.2 This requirement may be presumed to be met by the prequalified joint details of Figs. 10.13.1.1 and 10.13.1.2 when matching materials (Table 4.1.1) are used.

10.5.4 Laminations and Lamellar Tearing. Where tubular joints introduce through-thickness stresses, the anisotropy of the material and the possibility of base metal separation should be recognized during both design and fabrication.

10.6 Increased Unit Stresses

Where the applicable design specifications permit the use of increased unit stresses in the base metal for any reason, a corresponding increase shall be applied to the allowable unit stresses given herein, except for fatigue. The allowable stresses given herein are consistent with a nominal base metal working stress of $0.6F_y$.

10.7 Fatigue

10.7.1 Fatigue, as used herein, is defined as the damage that may result in fracture after a sufficient number of stress fluctuations. Stress range is defined as the peak-to-trough magnitude of these fluctuations. In the case of stress reversal, stress range shall be computed as the numerical sum (algebraic difference) of maximum repeated tensile and compressive stresses, or the sum of shearing stresses of opposite direction at a given point, resulting from changing conditions of load.

10.7.2 In the design of members and connections subject to repeated variations in live load stress, consideration shall be given to the number of stress cycles, the expected range of stress, and type and location of member or detail.

40. The ultimate breaking strength of fillet welds and partial joint penetration groove welds shall be computed at 2.67 times the basic allowable stress for 60 ksi (414 MPa) or 70 ksi (483 MPa) tensile strength and at 2.2 times the basic allowable stress for higher strength levels. The ultimate punching shear shall be taken as 1.8 times the allowable V_p of 10.5.1.

Table 10.7.3
Stress categories for type and location of material for circular sections

Stress category	Situation	Kinds of stress[1]
A	Plain unwelded pipe.	TCBR
A	Butt splices, no change in section, complete joint penetration groove welds, ground flush, and inspected by RT or UT.	TCBR
B	Pipe with longitudinal seam.	TCBR
B	Butt splices, complete joint penetration groove welds, ground flush.	TCBR
B	Members with continuously welded longitudinal stiffeners.	TCBR
C	Butt splices, complete joint penetration groove welds, as welded.	TCBR
D	Members with transverse (ring) stiffeners or miscellaneous attachments such as clips, brackets, etc.	TCBR
D	T- and cruciform joints with complete joint penetration welds (except at tubular connections).	TCBR
[2]D′	Connections designed as simple T-, Y-, or K-connections with complete joint penetration groove welds conforming to Fig. 10.13.1.1 (including overlapping connections in which the main member at each intersection meets punching shear requirements).	TCBR in branch member. (Note: Main member must be checked separately per category K or T.)
E	Balanced T- and cruciform joints with partial joint penetration groove welds or fillet welds (except at tubular connections).	TCBR in member; weld must also be checked per category G.
E	Members where doubler wrap, cover plates, longitudinal stiffeners, gusset plates, etc., terminate (except at tubular connections).	TCBR in member; weld must also be checked per category G.
[2]E′	Simple T-, Y-, and K-connections with partial joint penetration groove welds or fillet welds; also, complex tubular connections in which the punching shear capacity of the main member cannot carry the entire load and load transfer is accomplished by overlap (negative eccentricity), gusset plates, ring stiffeners, etc.	TCBR in branch member. (Note: Main member in simple T-, Y-, or K-connections must be checked separately per category K or T; weld must also be checked per category G′ and 10.5.3.)
F	End weld of cover plate or doubler wrap; welds on gusset plates, stiffeners, etc.	Shear in weld.
G	T- and cruciform joints, loaded in tension or bending, having fillet or partial joint penetration groove welds (except at tubular connections).	Shear in weld (regardless of direction of loading).
G′	Simple T-, Y-, or K-connections loaded in tension or bending, having fillet or partial joint penetration groove welds.	Shear in weld (regardless of direction of loading).

Table 10.7.3 (continued)
Stress categories for type and location of material for circular sections

Stress category	Situation	Kinds of stress[1]
X[6]	Intersecting members at simple T-, Y-, and K-connections; any connection whose adequacy is determined by testing an accurately scaled model or by theoretical analysis (e.g., finite element).	Total range of worst hot spot stress or strain on the outside surface of intersecting members at the tow of the weld joining them—measured after shakedown in model or prototype connection or calculated with best available theory.
X[6]	Unreinforced cone-cylinder intersection.	Hot-spot stress at angle change.[4]
K[3,6]	Simple K-connections in which gamma ratio R/t of main member does not exceed 24.	Punching shear on shear area of main member.[5]
T[3]	Simple T- and Y-connections in which gamma ratio R/t of main member does not exceed 24.	Punching shear on shear area of main member as defined in 10.8.

Notes to Table 10.7.3

1. T = tension, C = compression, B = bending, R = reversal—i.e., total range of nominal axial and bending stress.
2. Empirical curves (Fig. 10.7.4) based on "typical" connection geometries; if actual stress concentration factors or hot spot strains are known, use of curve X is preferred.
3. Empirical curves (Fig. 10.7.4) based on tests with gamma (R/t) of 18 to 24; curves on safe side for very heavy chord members (low R/t); for chord members (R/t greater than 24) reduce allowable stress in proportion to

$$\frac{\text{Allowable fatigue stress}}{\text{Stress from curve T or K}} = \left(\frac{24}{R/t}\right)^{0.7}$$

Where actual stress concentration factors or hot-spot strains are known, use of curve X is preferred.

4. Stress concentration factor—$SCF = \dfrac{1}{\cos\overline{\Psi}} + 1.17\sqrt{\gamma_b}\tan\overline{\Psi}$,
   where
   $\overline{\Psi}$ is angle change at transition
   γ_b is radius thickness ratio of tube at transition
5. Acting $V_p = \tau\sin\theta\left[f_a + \sqrt{2/3f_{by})^2 + (3/2f_{bz})^2}\right]$ where f_{by} and f_{bz} are nominal stresses for in-plane and out-of-plane bending, respectively.
6. For designated members which are controlled by fatigue calculations, a special effort should be made to achieve an as-welded surface which merges smoothly with the adjoining base metal and approximates the profile shown in Fig. 10.13.1.1(A).

10.7.3 The type and location of material shall be categorized as shown in Table 10.7.3.

10.7.4 Where the applicable design specification has a fatigue requirement, the maximum stress shall not exceed the basic allowable stress provided elsewhere, and the range of stress at a given number of cycles shall not exceed the values given in Fig. 10.7.4.

10.7.4.1 The increase in allowable stress provided in 10.6 shall not apply to the stress range values used in fatigue.

10.7.4.2 Where the fatigue environment involves stress ranges of varying magnitude and varying numbers of applications, the cumulative fatigue damage ratio, D, summed over all the various loads, shall not exceed unity,[41] where

$$D = \sum \frac{n}{N}$$

n = number of cycles applied at a given stress range
N = number of cycles for which the given stress range would be allowed in Fig. 10.7.4.

41. For critical members whose sole failure mode would be catastrophic, D should be limited to a fractional value (e.g., one-third.)

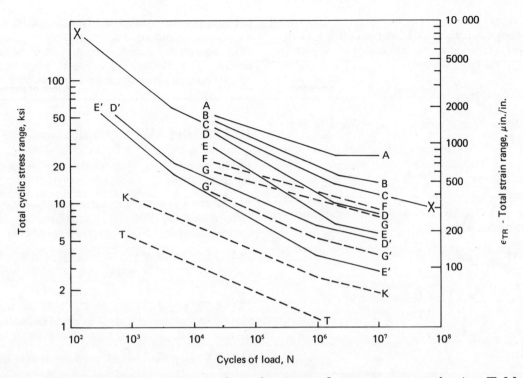

Fig. 10.7.4—Allowable fatigue stress and strain ranges for stress categories (see Table 10.7.3)

10.8 Effective Weld Areas, Lengths, and Throat

10.8.1 Groove Welds. The effective area shall be in accordance with 2.3.1 and the following: the effective length of groove welds in structural T-, K-, and Y-connections shall be computed in accordance with 10.8.4 or 10.8.5, using the mean radius or face dimensions of the branch member.

10.8.2 Fillet Welds. The effective area shall be in accordance with 2.3.2 and the following: the effective length of fillet welds in structural T-, Y-, and K-connections shall be computed in accordance with 10.8.4 or 10.8.5, using the radius or face dimensions of the branch member as measured to the center line of the weld.

10.8.3 Shear Area. The effective section for computing punching shear stress in simple T-, Y-, and K-connections shall be the main member thickness times the intersection length defined in 10.8.4 or 10.8.5, using the effective radius or face dimensions of the branch member as measured to the toe of the connecting weld on the main member outside surface.

10.8.4 Circular T-, Y-, K-Connections. Length of welds and the intersection length in circular T-, Y-, and K-connections shall be determined as $2\pi r K_a$ where

$$K_a = x + y + 3\sqrt{x^2 + y^2}$$

$$x = \frac{1}{2\pi \sin \theta}$$

$$y = \frac{1}{3\pi} \cdot \frac{3 - \beta^2}{2 - \beta^2}$$

θ = the acute angle between the two member axes
r = the effective radius of the intersection (see 10.8.3)
R = the outside radius of the main member
$\beta = \dfrac{r}{R}$

Note: The following may be used as conservative approximations:

$$K_a = \frac{1 + \dfrac{1}{\sin \theta}}{2} \quad \text{for axial load}$$

$$K_b = \frac{3 + \dfrac{1}{\sin \theta}}{4 \sin \theta} \quad \begin{array}{l}\text{for punching}\\ \text{shear or weld}\\ \text{stress due to}\\ \text{in-plane bending}\end{array}$$

$$K_b = \frac{1 + \dfrac{3}{\sin \theta}}{4 \sin \theta} \quad \begin{array}{l}\text{for punching}\\ \text{shear due to}\\ \text{out-of-plane}\\ \text{bending}\end{array}$$

$$K_b = \frac{1 + \dfrac{3}{\sin \theta}}{4} \quad \begin{array}{l}\text{for weld stress}\\ \text{due to out-of-}\\ \text{plane bending}\end{array}$$

10.8.5 Box Connections

10.8.5.1. Length of welds and intersection length in BOX CONNECTIONS shall be determined as K_a times the perimeter of the tube, where

$$K_a = \frac{a_x + b}{a + b} \quad \text{for axial load}$$

with a and b the face dimensions as defined in Fig. 10.1.2(B), and

$$a_x = \frac{a}{\sin \theta}$$

10.8.5.2 For bending, the section modulus of the actual intersection shall be taken as K_b times the section modulus of the tube, where

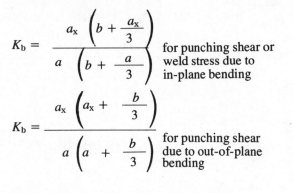

$$K_b = \frac{a_x \left(b + \dfrac{a_x}{3} \right)}{a \left(b + \dfrac{a}{3} \right)} \quad \text{for punching shear or weld stress due to in-plane bending}$$

$$K_b = \frac{a_x \left(a_x + \dfrac{b}{3} \right)}{a \left(a + \dfrac{b}{3} \right)} \quad \text{for punching shear due to out-of-plane bending}$$

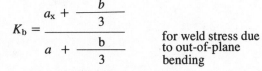

$$K_b = \frac{a_x + \dfrac{b}{3}}{a + \dfrac{b}{3}} \quad \text{for weld stress due to out-of-plane bending}$$

Part C
Structural Details

10.9 Combination of Welds

If two or more of the general types of welds (groove, fillet, plug, slot) are combined in a single joint, the allowable capacity of each shall be separately computed with reference to the axis of the group, in order to determine the allowable capacity of the combination.

10.10 Welds in Combination with Rivets and Bolts

Rivets or bolts used in bearing-type connections shall not be considered as sharing the stress in combination with welds. Welds, if used, shall be provided to carry the en-tire stress in the connection. However, connections that are welded to one member and riveted or bolted to the other member are permitted. High strength bolts (properly installed as a friction-type connection prior to welding) may be considered as sharing the stress with the welds.

10.11 Fillet Weld Details

10.11.1 Intermittent fillet welds may be used to carry calculated stress.

10.11.2 For lap joints, the minimum amount of lap shall be five times the thickness of the thinner part joined but not less than 1 in. (25.4 mm) (see Fig. 10.11.3).

10.11.3 Lap joints of telescoping tubes in which the load is transferred via the weld[42] may be single fillet welded in accordance with Fig. 10.11.3.

10.11.4 The maximum size of fillet weld that may be used along edges of material shall be equal to the thickness of the base metal.

10.11.5 Boxing shall be indicated on the drawings.

10.12 Transition of Thicknesses

Tension butt joints in axially aligned primary members of different material thicknesses or size shall be made in such a manner that the slope through the transition zone does not exceed 1 in 2-1/2. The transition shall be accomplished by chamfering the thick part, sloping the weld metal, or by any combination of these methods (see Fig. 10.12).

Part D
Details of Welded Joints

10.13 Prequalified Tubular Joints

10.13.1 Welded joints shall be in accordance with Part C, Section 2 of this code and the following paragraphs:

10.13.1.1 Complete Joint Penetration Tubular Groove Welds Made by Shielded Metal Arc, Gas Metal Arc,[43] or Flux Cored Arc Welding

(1) A complete joint penetration tubular groove weld made from one side only, without backing, is permitted where the size or configuration prevents access to the root side of the weld. Special skill is required for single-side welding of complete joint penetration tubular welds (see 5.21).

---

42. As opposed to an interference slip-on joint as used in tapered poles.

43. See Note (2) of Fig. 10.13.1.1(a) for restrictions in gas metal arc welding.

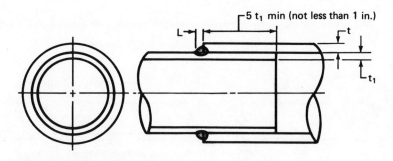

Note: L = size as required

Fig. 10.11.3—Fillet-welded lap joint

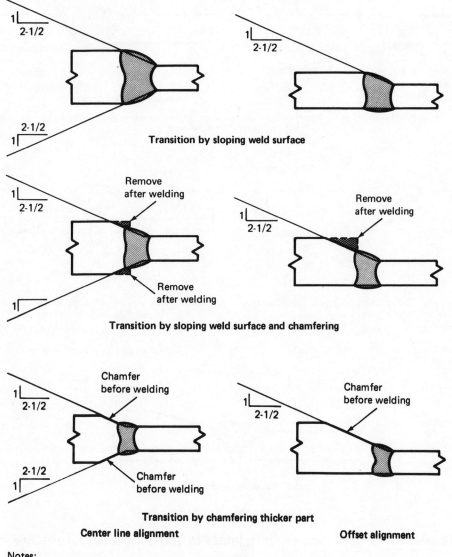

Transition by sloping weld surface

Transition by sloping weld surface and chamfering

Transition by chamfering thicker part

Center line alignment   Offset alignment

Notes:

1. Groove may be any permitted or qualified type and detail.
2. Transition slopes shown are the maximum permitted.

Fig. 10.12A—Transition of thickness of butt joints in parts of unequal thickness welded from two sides

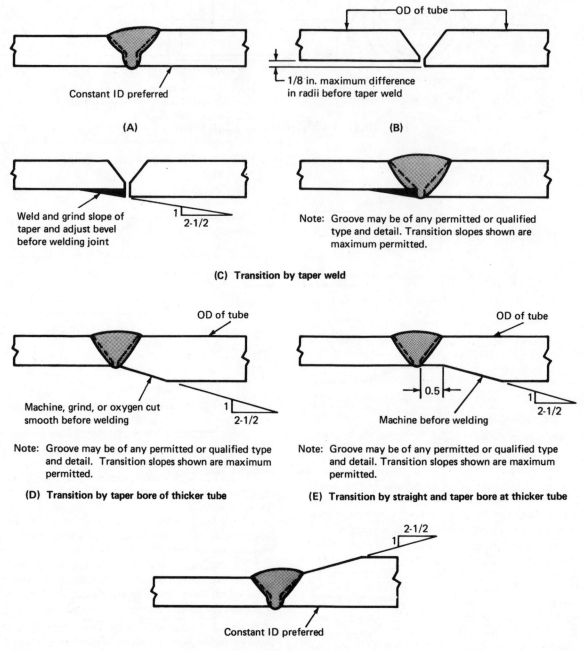

Constant ID preferred

(A)

OD of tube

1/8 in. maximum difference in radii before taper weld

(B)

Weld and grind slope of taper and adjust bevel before welding joint

1
2-1/2

Note: Groove may be of any permitted or qualified type and detail. Transition slopes shown are maximum permitted.

(C) Transition by taper weld

OD of tube

Machine, grind, or oxygen cut smooth before welding

1
2-1/2

Note: Groove may be of any permitted or qualified type and detail. Transition slopes shown are maximum permitted.

(D) Transition by taper bore of thicker tube

OD of tube

0.5

Machine before welding

1
2-1/2

Note: Groove may be of any permitted or qualified type and detail. Transition slopes shown are maximum permitted.

(E) Transition by straight and taper bore at thicker tube

2-1/2
1

Constant ID preferred

(F) Transition by taper OD of thicker tube

Fig. 10.12B—Transition of thickness of butt joints in parts of unequal thickness welded from one side

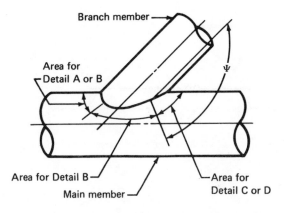

Branch member

Area for
Detail A or B

Ψ

Area for Detail B

Main member

Area for
Detail C or D

Notes:

1. The applicable joint detail (A, B, C, or D) for a particular part of the connection is determined by the local dihedral angle, Ψ, which changes continuously in progressing around the branch member, as follows:

Detail	Applicable range of local dihedral angle
A	180° to 135°
B	150° to 50°
C	75° to 30° (not prequalified for groove angles under 30°)
D	40° to 15° (not prequalified for groove angles under 30°)

2. The angle and dimensional ranges given in Detail A, B, C, or D include maximum allowable tolerances.

Fig. 10.13.1.1A—Complete joint penetration prequalified circular-tubular joints for simple T-, Y-, or K-connections made by shielded metal arc, gas metal arc, or flux cored arc welding

(2) Complete joint penetration tubular groove welds made by shielded metal arc, gas metal arc,[43] or flux cored arc welding, that may be used without performing joint welding procedure qualification tests prescribed by 5.2 are detailed in Fig. 10.13.1.1 (A) and (B) and are subject to the limitations specified in Fig. 10.13.1.1 and in Table 10.13.1.1.

(3) The weld joint angles and dimensions shall not deviate from the ranges detailed in Fig. 10.13.1.1. The root face of the joints is zero unless dimensioned otherwise. It may be detailed to exceed zero or the specified dimension by not more than 1/16 in. (1.6 mm). It may not be detailed to less than the specified dimension.

(4) The sample joints or mock-up shall provide at least one macroetched test section for each of the following conditions:

(a) The groove combining the greatest groove depth with the smallest groove angle (or combinations

---
43. See Note 2A of Fig. 10.13.1.1 for restrictions in gas metal arc welding.

Table 10.13.1.1
Limitations on prequalified tubular joint welding procedures

	SMAW	GMAW*	FCAW	New procedure or modified joint details
Test for metallurgical compatibility and mechanical properties	None, if sections 3 and 4 are met	6G or 2G+5G	None, if sections 3 and 4 are met	6G, or 2G+5G
Sample joints, see 10.13.1.1 (4), or full scale mock-up to check suitability of joint details	None	None	Required	Required
Required level of welder skill	6GR	6GR	6GR	Prior 6GR test acceptable

*See Note 2A of Fig. 10.13.1.1 for restrictions in gas metal arc welding.

of grooves) to be used; test with the welding position vertical.

(b) The narrowest root opening to be used with a 37.5 deg groove angle; one test each with the welding position flat and overhead.

(c) The widest root opening to be used with a 37.5 deg groove angle; one test each with the welding position flat and overhead.

(d) For matched box connections only, the minimum groove angle and corner radius to be used in combination; one test in the horizontal position.

(5) The macroetched test specimens required in (4) shall be examined for discontinuities, and those specimens which have discontinuities prohibited by 10.17, as applicable, shall be considered as failed (see 5.12.3).

10.13.1.2 Complete Joint Penetration Tubular Groove Welds in Butt Joints Made by Shielded Metal Arc, Gas Metal Arc,[43] or Flux Cored Arc Welding

(1) Where welding from both sides or welding from one side with backing is possible, any procedure and groove detail that is appropriately prequalified in accordance with 5.1 may be used. Other procedures and groove details qualified in accordance with 5.2 may be used.

(2) Butt joints welded from one side, without backing, shall not be considered as complete joint penetration

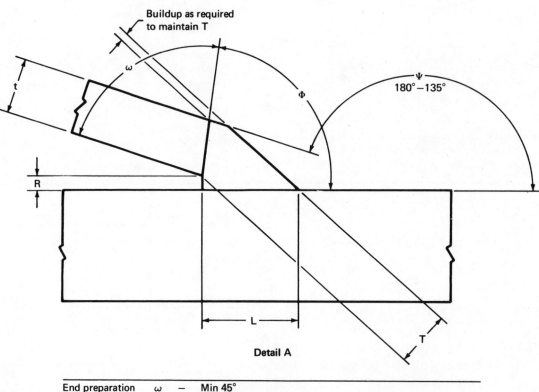

Detail A

End preparation	ω	—	Min 45°		
			Max 90° (square cut)		
Fit-up			For SMAW[1]		For GMAW and FCAW[2]
	R	—	Max 3/16 in.	R —	Max 3/16 in.
			Min 1/16 in.		Min 1/16 in.
			(No min for Φ over 90°)		(No min for Φ over 120°)
Completed weld	T	—	Shall be not less than t		
	L	—	Shall be not less than t/sin Ψ but need not exceed 1.75t		

Notes:

1. Detail for SMAW also applies to FCAW and GMAW for which the root pass is made by SMAW.
2. Detail for GMAW and FCAW applies to (1) GMAW short-circuiting transfer procedures qualified in accordance with 5.2 and (2) FCAW procedures which have been tested according to 10.13.1.1(4) and 10.13.1.1(5).
3. t = branch member thickness.

Fig. 10.13.1.1A (continued)—Complete joint penetration prequalified circular-tubular joints for simple T-, Y-, or K-connections made by shielded metal arc, gas metal arc, or flux cored arc welding

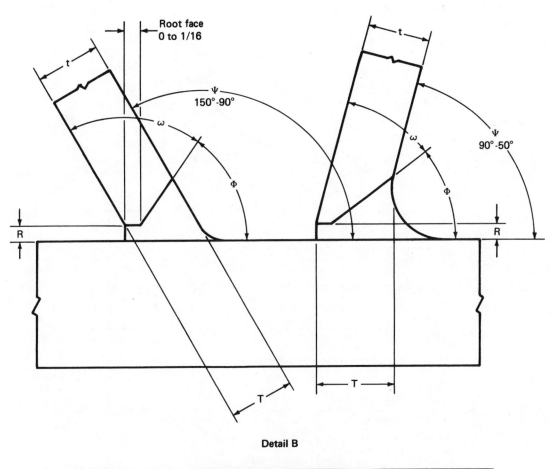

Detail B

End preparation	ω	—	Min 10° (45° for Ψ over 105°)
			Max 90° (square cut)
			Otherwise as needed to obtain required Φ

Fit-up	Φ	—	Min 37.5° (if less, use Detail C)		
			Max 60° for Ψ up to 105°		
			For SMAW[1]		For GMAW and FCAW[2]
	R	—	Min 1/16 in.	R —	Min 1/16 in.
			Max 1/4 in.		Max 1/4 in. for Φ over 45°
					Max 5/16 in. for Φ not over 45°

Completed weld	T	—	Shall be not less than t for Ψ over 90°
	T	—	Shall not be less than t/sin Ψ for Ψ not over 90°

Notes:

1. Detail for SMAW also applies to FCAW and GMAW for which the root pass is made by SMAW.
2. Detail for GMAW and FCAW applies to (1) GMAW short-circuiting transfer procedures qualified in accordance with 5.2 and (2) FCAW procedures which have been tested according to 10.13.1.1(4) and 10.13.1.1(5).
3. t = branch member thickness.

Fig. 10.13.1.1A (continued)—Complete joint penetration prequalified circular-tubular joints for simple T-, Y-, or K-connections made by shielded metal arc, gas metal arc, or flux cored arc welding

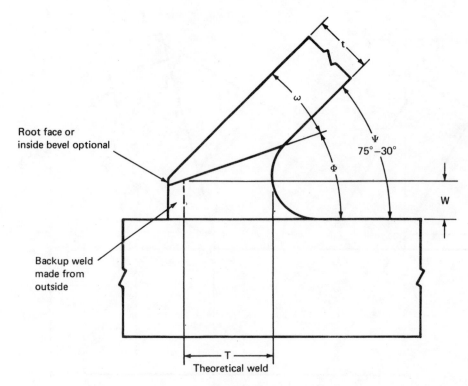

Detail C

End preparation	ω	—	As needed to obtain required Φ min 10° (approximately 1:6 bevel)
Fit-up	Φ	—	Min $\Psi/2$ Max 37.5° (if more, use Detail B) Root opening not to exceed W tabulated below
Completed weld	W	—	Backup weld—initial passes discounted until width of groove W is sufficient to assure sound welding (see table below)
	T	—	Theoretical weld -- not less than t/sin Ψ but need not exceed 1.75t (weld may be built up to meet this requirement)

For SMAW[1]

W	—	1/8 in.	3/16 in.	
Φ	—	22.5° to 37.5°	15° to 22.5°	

For GMAW and FCAW[2, 3]

W	—	1/8 in.	1/4 in.	3/8 in.	1/2 in.
Φ	—	30° to 37.5°	25° to 30°	20° to 25°	15° to 20°

Notes:
1. Detail for SMAW also applies to FCAW and GMAW for which the root pass is made by SMAW.
2. Detail for GMAW and FCAW applies to (1) GMAW short-circuiting transfer procedures qualified in accordance with 5.2 and (2) FCAW procedures which have been tested according to 10.13.1.1(4) and 10.13.1.1(5).
3. In Details C and D, backup weld may be made by SMAW.
4. t = branch member thickness.

Fig. 10.13.1.1A (continued)—Complete joint penetration prequalified circular-tubular joints for simple T-, Y-, or K-connections made by shielded metal arc, gas metal arc, or flux cored arc welding

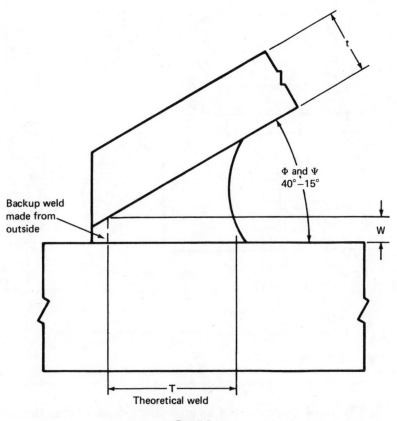

Detail D

End preparation	ω	—	No outside bevel; inside bevel optional; transition from Detail C is to be gradual
Fit-up	Φ	—	Root opening not to exceed W tabulated below
Completed weld	W	—	Backup weld—initial passes discounted until width of groove W is sufficient to assure sound welding (see table below)
	T	—	Theoretical weld — not less than 2t

For SMAW[1]

W	—	1/8 in.	3/16 in.
Φ	—	22.5° to 37.5°	15° to 22.5°

For GMAW and FCAW[2, 3]

W	—	1/8 in.	1/4 in	3/8 in.	1/2 in.
Φ	—	30° to 37.5°	25° to 30°	20° to 25°	15° to 20°

Notes:
1. Detail for SMAW also applies to FCAW and GMAW for which the root pass is made by SMAW.
2. Detail for GMAW and FCAW applies to (1) GMAW short-circuiting transfer procedures qualified in accordance with 5.2 and (2) FCAW procedures which have been tested according to 10.13.1.1(4) and 10.13.1.1(5).
3. In Details C and D, backup weld may be made by SMAW.
4. t = branch member thickness.

Fig. 10.13.1.1A (continued)—Complete joint penetration prequalified circular-tubular joints for simple T-, Y-, or K-connections made by shielded metal arc, gas metal arc, or flux cored arc welding.

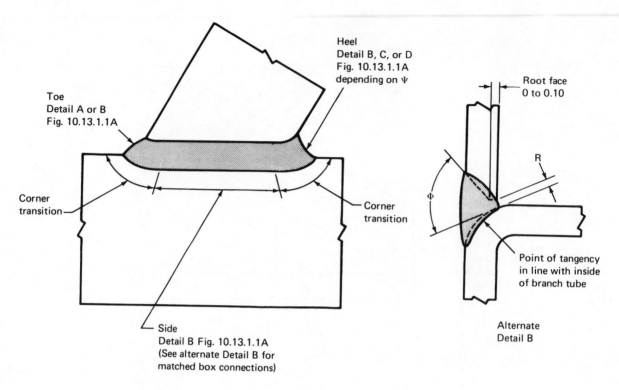

Notes:

1. Details A, B, C, and D and all notes from Fig. 10.13.1.1A apply.
2. Joint preparation for corner welds shall provide a smooth transition from one detail to another. Welding shall be carried continuously around corners, with corners fully built up and all starts and stops within flat faces.

Fig. 10.13.1.1B—Complete joint penetration prequalified box connections made by shielded metal arc, gas metal arc, or flux cored arc welding

groove welds, unless all of the following provisions are complied with:

    (a) The joint detail and welding procedure are qualified by appropriate tests in accordance with 5.2.

    (b) The welders have been qualified to weld pipe or tubing, without backing, in accordance with 5.17.4.

    (c) If the groove design to be used varies from Fig. 5.20A, the actual groove design used in construction shall be used for the tests required by (a) and (b).

    (d) All completed welds shall be 100% examined nondestructively by either radiographic testing or ultrasonic testing and the weld quality shall conform to 10.17.2.

10.13.1.3 Partial joint penetration circular-tubular groove welds made by shielded metal arc, gas metal arc,[43] and flux cored arc welding, that may be used without performing the joint welding procedure qualification tests prescribed in 5.2, are shown in Figs. 10.13.1.3 (A) and (B).

10.13.1.4 For partial joint penetration groove welds in matched box connections, if the corner radius of the main tube is less than two times the thickness of the branch tube, a simple joint of the side detail shall be made and sectioned to verify the required effective throat. The test weld shall be made in the horizontal position.

10.13.1.5 Fillet welded circular-tubular connections made by shielded metal arc, gas metal arc,[43] and flux cored arc welding, that may be used without performing the joint welding procedure qualification tests prescribed by 5.2, are detailed in Part C, Section 2 of this Code and in Fig. 10.13.1.5.

Note: Prequalified fillet weld detail limited to $\beta \le 1/3$ for circular and $\beta \le 0.8$ for box connections.

Part E
Workmanship

10.14 Assembly

10.14.1 The parts to be joined by fillet welds shall be brought into the closest practicable contact, and in no event shall be separated by more than 3/16 in. (4.8 mm). If the separation is 1/16 in. (1.6 mm) or greater, the leg of the fillet weld shall be increased by the amount of the

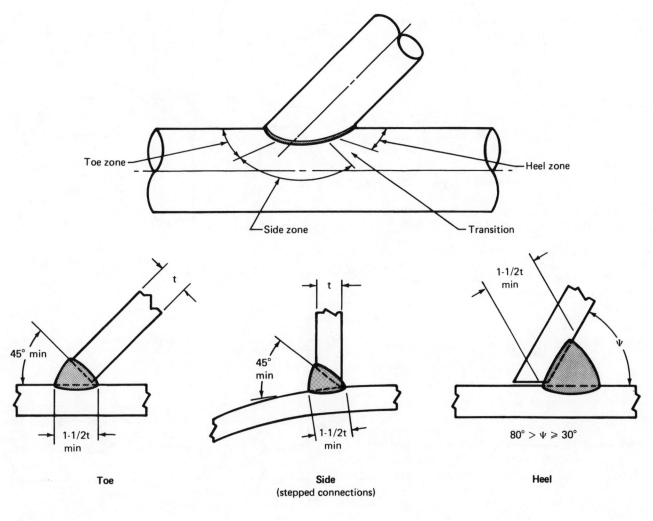

Notes:

1. t = thickness of thinner section
2. Depth of bevel = t
3. Root opening 0 to 3/16 in. (4.8 mm)
4. Ψ = 15 ° min (not prequalified for under 30°)
5. Effective throat = t

Fig. 10.13.1.3A—Partial joint penetration prequalified circular-tubular joints made by shielded metal arc, gas metal arc, or flux cored arc welding

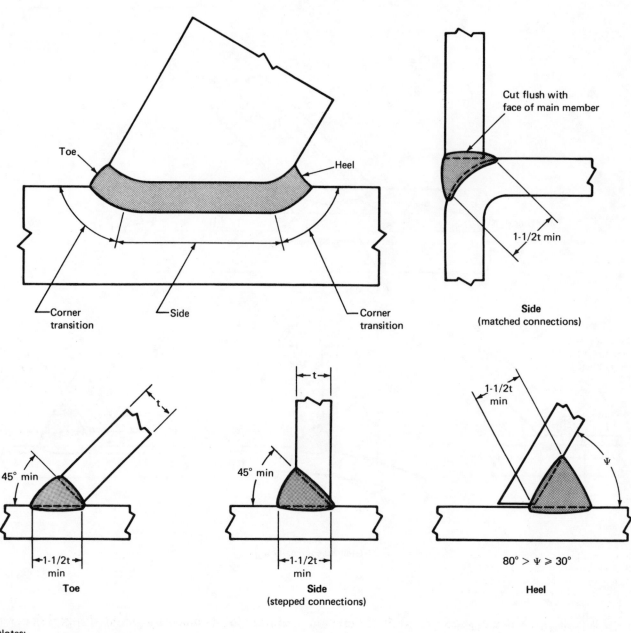

Notes:

1. t = thickness of thinner section
2. Depth of bevel = t
3. Root opening 0 to 3/16 in. (4.8 mm)
4. Not prequalified for Ψ under 30°
5. Effective throat = t
6. Joint preparation for corner welds shall provide a smooth transition from one detail to another. Welding shall be carried continuously around corners, with corners fully built up and all starts and stops within flat faces.

Fig. 10.13.1.3B—Partial joint penetration prequalified box connections made by shielded metal arc, gas metal arc, or flux cored arc welding

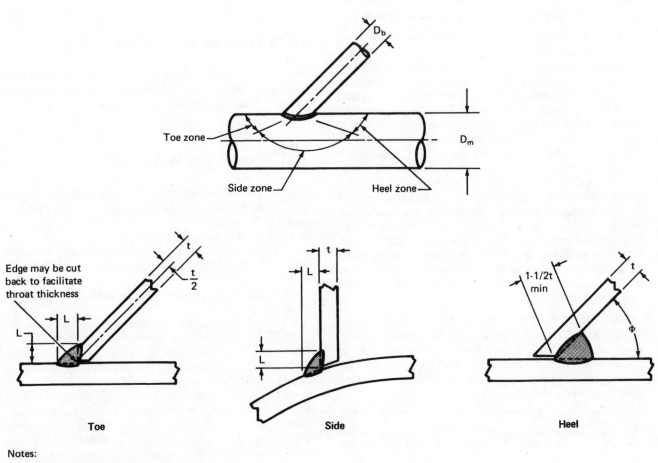

Notes:

1. t = thickness of thinner part
2. L = minimum size = t (see 10.5.3)
3. Root opening 0 to 3/16 in. (4.8 mm)
4. Φ = 15° min (not prequalified for under 30°)
5. See 10.13.1.5 for limitations on $\beta = D_b/D_m$

Fig. 10.13.1.5—Fillet welded prequalified circular-tubular joints made by shielded metal arc, gas metal arc, and flux cored arc welding

separation. The separation between faying surfaces of lap joints and of butt joints landing on a backing shall not exceed 1/16 in. (1.6 mm). Where irregularities in rolled shapes, after straightening, do not permit contact within the above limits, the procedure necessary to bring the material within these limits shall be subject to the approval of the Engineer. The use of fillers is prohibited except as specified on the drawings or as specially approved by the Engineer and made in accordance with 2.4.

10.14.2 Abutting parts to be joined by girth welds shall be carefully aligned. No two girth welds shall be located closer than one pipe diameter or 3 ft (900 mm), whichever is less. There shall be no more than two girth welds in any 10 ft (3 m) interval of pipe except as may be agreed upon by the owner and contractor. Radial offset of abutting edges of girth seams shall not exceed 0.2t (where t is the wall thickness) and the maximum allowable shall be 1/4 in. (6.4 mm) provided that any offset exceeding 1/8 in. (3.2 mm) is welded from both sides. However, with the approval of the Engineer, one localized area per girth seam may be offset up to 0.3t with a maximum of 3/8 in. (9.5 mm), provided the localized area is under 8t in length and that weld metal is added to this region to provide a 4 to 1 transition. Offsets in excess of this shall be corrected as provided in 3.3.3. Longitudinal weld seams of adjoining sections shall be staggered a minimum of 90 deg unless closer spacing is agreed upon by the owner and fabricator.

10.14.3 Variation in cross section dimensions of groove welded joints, from those shown on the detail drawings, shall be in accordance with those shown in 3.3.4. In addition, the following tolerances will apply to complete joint penetration tubular groove welds made from one side only without backing:

	Root face of joint		Root opening of joints without steel backing[44]		Groove angle of joint
	in	mm	in.	mm	deg
SMAW	±1/16	1.6	±1/16	1.6	±5
GMAW	±1/32	0.8	±1/16	1.6	±5
FCAW	±1/16	1.6	±1/16	1.6	±5

44. Root openings wider than permitted by the above tolerances but not greater than the thickness of the thinner part may be built up by welding to acceptable dimensions prior to the joining of the parts by welding.

10.15 Temporary Welds

Temporary welds shall be subject to the same welding procedure requirements as the final welds. They shall be removed unless otherwise permitted by the Engineer. When they are removed, the surface shall be made flush with the original surface. There shall be no temporary welds in tension zones of members made of quenched and tempered material. Temporary welds at other locations shall be shown on shop drawings and shall be made with E70XX low-hydrogen electrodes.

10.16 Dimensional Tolerances

The dimensions of tubular members shall be within the tolerances specified in 3.5 wherein the term column is interpreted as compression tubular member.

10.17 Quality of Welds

10.17.1 Visual Inspection. All welds shall be visually inspected. A weld shall be acceptable by visual inspection if it shows that

10.17.1.1 The weld has no cracks.

10.17.1.2 Thorough fusion exists between adjacent layers of weld metal and between weld metal and base metal.

10.17.1.3 All craters are filled to the full cross section of the weld.

10.17.1.4 Weld profiles are in accordance with 3.6.

10.17.1.5 The sum of diameters of piping porosity in fillet welds does not exceed 3/8 in. (9.5 mm) in any linear inch of weld and does not exceed 3/4 in. (19.0 mm) in any 12 in. (305 mm) length of weld.

10.17.1.6 Complete joint penetration groove welds in butt joints transverse to the direction of computed tensile stress shall have no piping porosity. For all other groove welds, piping porosity shall not exceed 3/8 in. (9.5 mm) in any linear inch of weld and shall not exceed 3/4 in. (19 mm) in any 12 in. (305 mm) length of weld.

10.17.1.7 Visual inspection of welds in all steels may begin immediately after the completed welds have cooled to ambient temperature. Acceptance criteria for ASTM A514 and A517 steels shall be based on visual inspection performed not less than 48 hours after completion of the weld.

10.17.2 Radiographic and Magnetic Particle Inspection. Welds that are subject to radiographic or magnetic particle testing, in addition to visual inspection, shall have no cracks and shall be unacceptable if the radiographic or magnetic particle inspection shows discontinuities that exceed the following limitations:

10.17.2.1 Individual discontinuities, having a greatest dimension of 3/32 in. (2.4 mm) or greater, if

(1) The greatest dimension of a discontinuity is larger than 2/3 of the effective throat, 2/3 the weld size, or 3/4 in. (19.0 mm).

(2) The discontinuity is closer than three times its greatest dimension to the end of a groove weld subject to primary tensile stresses.

(3) A group of such discontinuities is in line such that

(a) The sum of the greatest dimensions of all such discontinuities is larger than the effective throat or weld size in any length of six times the effective throat or weld size. When the length of the weld being examined is less than six times the effective throat or weld size, the permissible sum of the greatest dimensions shall be proportionally less than the effective throat or weld size.

(b) The space between two such discontinuities which are adjacent is less than three times the greatest dimension of the larger of the discontinuities in the pair being considered.

10.17.2.2 Independent of the requirements of 10.17.2.1, discontinuities having a greatest dimension of less than 3/32 in. (2.4 mm) if the sum of their greatest dimensions exceeds 3/8 in. (9.5 mm) in any linear inch of weld.

10.17.3 Liquid Penetrant Inspection. Welds that are subject to liquid penetrant testing, in addition to visual inspection, shall be evaluated on the basis of the requirements for visual inspection.

10.17.4 Ultrasonic Inspection. When ultrasonic testing is required, the testing procedure and acceptance criteria shall be specified in the contract.

10.17.5 When welds are subject to nondestructive testing in accordance with 10.17.2, 10.17.3, and 10.17.4, the testing may begin immediately after the completed welds have cooled to ambient temperature. Acceptance criteria for ASTM A514 and A517 steels shall be based on nondestructive testing performed not less than 48 hours after completion of the welds.

Appendix A
Effective Throat

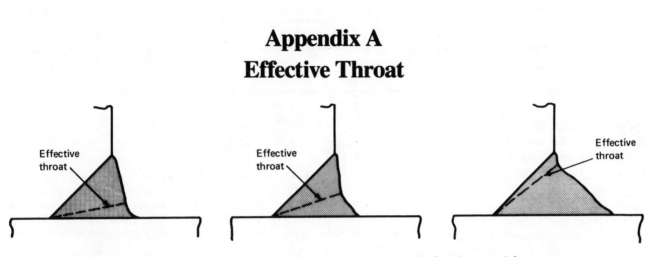

Note: The effective throat of a weld is the minimum distance from the root of a weld to its face, less any reinforcement.

Appendix B
Effective Throats of Fillet Welds in Skewed T-Joints

Table B is a tabulation showing equivalent leg size factors for the range of dihedral angles between 60 deg and 135 deg, assuming no gap. Gaps 1/16 in. (1.6 mm) or greater, but not exceeding 3/16 in. (5 mm), shall be added directly to the leg size. The required leg size for fillet welds in skewed joints is calculated using the equivalent leg size factor for correct dihedral angle, as shown in the example.

EXAMPLE
(U.S. customary units)

Given: Skewed T-joint, angle: 75 deg; gap: 1/16 (0.063) in.

Required: Strength equivalent to 90 deg fillet weld of size: 5/16 (0.313) in.

Procedure: (1) Factor for 75 deg from Table B: 0.86

(2) Equivalent leg size, w, of skewed joint, without gap:
w = 0.86 × 0.313 = 0.269 in.

(3) With gap of: 0.063 in.

(4) Required leg size, w, of skewed fillet weld: [(2) + (3)] w = 0.332 in.

(5) Rounding up to a practical dimension: w = 3/8 in.

EXAMPLE
(SI units)

Given: Skewed T-joint, angle: 75 deg; gap: 1.6 mm

Required: Strength equivalent to 90 deg fillet weld of size: 8.0 mm

Procedure: (1) Factor for 75 deg from Table B: 0.86

(2) Equivalent leg size, w, of skewed joint, without gap:
w = 0.86 x 8.0 = 6.9 mm

(3) With gap of: 1.6 mm

(4) Required leg size, w, of skewed fillet weld: [(2) + (3)] 8.5 mm

(5) Rounding up to a practical dimension: w = 9.0 mm

175

Table B
Equivalent fillet weld leg size factors for skewed T-joints

Dihedral angle, ψ	60	65	70	75	80	85	90	95
Comparable fillet weld size for same strength	0.71	0.76	0.81	0.86	0.91	0.96	1.00	1.03
Dihedral angle, ψ	100	105	110	115	120	125	130	135
Comparable fillet weld size for same strength	1.08	1.12	1.16	1.19	1.23	1.25	1.28	1.31

For fillet welds having equal measured legs (w_n), the distance from the root of the joint to the face of the diagrammatic weld (t_n) may be calculated as follows:

For gaps $>$ 1/16 in. and $\leq$ 3/16 in., use

$$t_n = \frac{w_n - g_n}{2 \sin \dfrac{\psi}{2}}$$

For gaps $<$ 1/16 in., use
$$g_n = 0 \text{ and } t'_n = t_n$$

where the measured leg of such fillet weld (w_n) is the perpendicular distance from the surface of the joint to the opposite toe, and (g) is the gap, if any, between parts. See Fig. 2.7.1. Acceptable gaps are defined in 3.3.1.

Appendix C
Impact Strength Requirements for
Electroslag and Electrogas Welding

C1 General

The impact test requirements and test procedures in this appendix shall apply only when specified in the contract drawings or specifications in accordance with 4.15.3 of this code.

C2 Impact Properties

C2.1 When computing the average value of the impact properties, the extreme lowest value and extreme highest value obtained with the five specimens shall be disregarded.

C2.2 The notched bar impact properties of the weld metal shall be no less than the values in Table C2.2 when tested at $0°$ F $(-18°$ C$)$.

C2.3 If the value for more than one of the three specimens is below the minimum average requirement, or if the value for one of the three specimens is below the minimum value permitted on one specimen, a retest shall be made, and the value of all three specimens must equal or exceed the specified minimum average value. Such a retest shall be permitted only when the average value of the three specimens equals or exceeds the minimum value permitted on one specimen.

C3 Impact Test

C3.1 Five Charpy V-notch impact test specimens shall be machined from the same test weld assembly (Fig. 5.10.1.3C) made to determine weld joint properties.

C3.2 The impact specimens shall be machined and tested in accordance with ASTM E23, Standard Methods for Notched Bar Impact Testing of Metallic Materials, for Type A Charpy (simple beam) impact specimen.

C3.3 The longitudinal center line of the specimens shall be transverse to the weld axis, and for material of thickness greater than 1/2 in. (12.7 mm) shall be located as near as practicable to a point midway between the surface and the center of thickness. The base of the notch shall be perpendicular (normal) to the surface. The standard 10 mm × 10 mm specimen shall be used where the thickness is 1/2 in. or greater, and for thinner material the largest possible subsize specimen listed in Table C2.2 shall be used.

Table C2.2
Impact strength requirements for electroslag and electrogas welding

Size of specimen	Minimum impact value required for average of three specimens		Minimum impact value permitted on one specimen only in a set of three specimens	
	ft-lb	J	ft-lb	J
10.0 mm × 10.0 mm	15.0	20.3	10.0	13.6
10.0 mm × 7.5 mm	12.5	17.0	8.5	11.5
10.0 mm × 5.0 mm	10.0	13.6	7.0	9.5
10.0 mm × 2.5 mm	5.0	6.8	3.5	4.7

Appendix D
Short Circuiting Transfer

Short circuiting transfer is a type of metal transfer in gas metal arc welding in which melted material from a consumable electrode is deposited during repeated short circuits. For additional information, see Volume 2 of the Seventh Edition of the Welding Handbook, page 115 and page 117, Table 4.1.

Typical current ranges for
short-circuiting transfer gas metal arc welding of steel

Electrode diameter		Welding current, amperes*			
		Flat position		Vertical and overhead positions	
in.	mm	min	max	min	max
0.030	0.8	50	150	50	125
0.035	0.9	75	175	75	150
0.045	1.2	100	225	100	175

*Electrode positive

Appendix E

Part A
Joint Welding Procedure Requirements

This part includes two tables and check lists for use in preparing either Form E-1, Prequalified Joint Welding Procedure, or Form E-2, Welding Procedure Qualification Test Record. Table E1 covers the mandatory requirements for according prequalified status to joint welding procedures (see 5.1.1). Table E2 covers the provisions of the code that may be modified when the joint welding procedure is qualified by tests (see 5.2). The check list is for use in preparing prequalified joint welding procedures.

Table E1
Mandatory code requirements for prequalified joint welding procedures

Note: All of the applicable provisions listed below must be complied with for a joint welding procedure to have prequalified status. All other code requirements are mandatory also, but the provisions listed below apply specifically to prequalified joint welding procedures.

Code provision	Subject
1.2	Base Metal
1.3	Welding Processes
Sect. 2, Part C Sect. 10, Part D	Details of Welded Joints
3.2.1 & 3.11	Weld Cleaning
3.3.4	Tolerance for dimensions of cross section of groove joints
3.8	Peening
4.1	Filler Metal Requirements
4.2	Preheat and Interpass Temperature Requirements
4.3	Heat Input Control for Quenched and Tempered Steel
4.5	Electrodes for SMAW[1]
4.6	Procedures for SMAW[1]
4.7.3	Maximum diameter of electrodes for SAW[2]
4.7.4	Cleaning—see also 3.2.1 and 3.11[2]
4.7.7	Cross section of groove or fillet weld[2]
4.8.1	Electrodes and flux for SAW[2]
4.9	Condition of Flux[2]
4.10	Procedures for SAW with Single Electrode[2]
4.11	Procedures for SAW with Multiple Electrodes[2]
4.12	Electrodes for GMAW and FCAW[3]
4.13	Shielding Gas[3]
4.14	Procedures for GMAW and FCAW with Single Electrode[3] (Note: GMAW and FCAW with multiple electrodes, GMAW-S, EGW, and ESW do not have prequalified status.)
Sect. 4, Part F	Stud Welding[4]
Sect. 4, Part G	Plug and Slot Welds (see also 2.8)[5]
5.1.2	Written procedure specification must be prepared
8.2.1 8.2.2 8.2.4	Base metal for buildings[6]
9.2.1 9.2.2 9.2.5	Base metal for bridges[6]
9.12	Prohibited Types of Joints and Welds[7]
10.2.1 10.2.2 10.2.4	Base metal for tubular structures[6]

1. Applicable only to shielded metal arc welding procedures.
2. Applicable only to submerged arc welding procedures.
3. Applicable only to gas metal arc and flux cored arc welding procedures.
4. Applicable only to stud welding.
5. Applicable only to plug and slot welds.
6. Type of structure governs the base metal to be used.
7. Applicable only to bridges or, when specified, other fatigue-loaded structures.

Table E2
Code requirements that may be changed by procedure qualification tests

Notes:
1. The code provisions listed below may be modified, changed, or disregarded when the joint welding procedure is established by tests (see 5.2 and Section 5, Part B), provided that in preparing the welding procedure specification specific values for each essential variable for the welding process listed in 5.5 are addressed, and any change of essential variables in 5.5.2 shall be within prescribed limits.

2. No other code requirements (not listed in Table E2) may be changed when the procedure is established by tests.

Code provision	Subject
1.2	Base Metal
1.3.4	Welding Processes
Sec. 2, Part C Sec. 10, Part D	Details of Welded Joints
3.11	Weld Cleaning
4.1	Filler Metal Requirements
4.2	Preheat and Interpass Temperature Requirements[1]
4.5.1, 4.5.4	Electrodes for SMAW
4.6	Procedures for SMAW
4.7.3	Maximum diameter of electrode for SAW
4.7.7	Cross section of SAW groove or fillet weld
4.8.1	Electrodes and flux for SAW
4.10	Procedures for SAW with Single Electrode
4.11	Procedures for SAW with Multiple Electrodes
4.12	Electrodes for GMAW and FCAW
4.14.1.2 4.14.1.3 4.14.1.4 4.14.1.5 4.14.2	Procedures for GMAW and FCAW with single electrode (Note: GMAW and FCAW with multiple electrodes, GMAW-S, EGW, and ESW do not have prequalified status.)
8.2.1 8.2.4	Base metal for buildings
9.2.1 9.2.5	Base metal for bridges
10.2.1 10.2.4	Base metal for tubular structures

1. The preheat and interpass temperature used during the procedure qualification is applicable only to that material thickness. When the material thickness to be welded in construction is other than the thickness used in qualification, the following provisions shall be observed: (1) When the steel is listed in 10.2, the requirements of Table 4.2 shall apply; (2) for other steels, the procedure specification shall include preheat and interpass temperatures applicable to varying material thickness in increments similar to those in Table 4.2.

Check list for prequalified joint welding procedures

This check list is provided for the convenience of those who have the responsibility of preparing prequalified joint welding procedures in accordance with 5.1.1 and 5.1.2. The provisions listed below relate directly to the information that is required by 5.1.1 and 5.1.2.

Part A lists requirements that must be met by all processes. Part B lists requirements applicable to the welding process to be used. When a combination of processes is used (see 5.1.3), requirements for each process must also be observed.

Welding procedure no._____ Date_____

Revision no._____

General requirements—all processes

Procedure variable	Applicable provision	Indicate compliance
Joint details	Figs. 2.7.1, 2.9.1, and 2.10.1. 2.8 , 4.24.11.1 (Studs only)	_____
Joint detail tolerances		
Fillet welds	2.7.1.1	_____
	2.7.1.2	_____
	2.7.1.4	_____
	2.7.1.5	_____
	2.7.1.6	_____
Plug and slot welds	2.8.2	_____
	2.8.3	_____
	2.8.4	_____
	2.8.5	_____
	2.8.6	_____
	2.8.7	_____
	2.8.8	_____

	Complete penetration	Partial penetration	
Groove welds	2.9.2.2	2.10.2.1	_____
	2.9.2.3	2.10.2.2	_____
	2.9.2.4	2.10.2.3	_____
	2.9.2.5	2.10.2.4	_____
	2.9.2.6	2.10.3	_____
		2.10.4	_____
Position (as permitted by applicable joint detail)	Figs. 2.9.1, 2.10.1		_____
Base metal thickness	2.9.2.1		_____
Minimum effective throat (partial penetration joints)	2.10.3		_____

General requirements—all processes (continued)

Procedure variable	Applicable provision	Indicate compliance
General requirements	3.1.3	_____
	3.2.1	_____
	3.3.1 (Fillets)	_____
	3.3.2 (Partial penetration only)	_____
	3.3.3	_____
	3.3.4	_____
	3.3.5	_____
	3.3.7	_____
	3.4	_____
	3.6.1	_____
	3.6.2	_____
	3.6.3	_____
	3.6.4	_____
	3.6.5	_____
	3.6.6	_____
	3.8	_____
	3.9	_____
	3.10	_____
	3.11	_____
	3.12	_____
	3.13	_____
	4.1.3	_____
	4.3	_____
Preheat and interpass temperature	4.2, 4.24.11.5 (Studs only)	_____
Postheat treatment	4.4	_____
Base metal type/grade	8.2, 9.2, 10.2	_____

Specific requirements applicable to welding process

Note: The following requirements apply only to the welding process being used.

Shielded metal arc welding

Procedure variable	Applicable provision	Indicate compliance
Filler metal: Specification classification	4.1.1, 4.1.4, 4.5.1 Tables 4.1.1, 4.1.4	_____ _____
Single or multiple pass	4.1.5.1	_____
General requirement	4.1.3 4.5 4.6.8	_____ _____ _____
Electrode (low hydrogen storage)	4.5.2, 4.5.3, 4.5.4	_____ _____
Welding current	4.6.2	_____
Electrode restrictions	4.6.3.1 4.6.3.2 4.6.3.3 4.6.3.4 4.6.3.5	_____ _____ _____ _____ _____
Root pass and weld layer restrictions	4.6.4 4.6.5 4.6.6.1 4.6.6.2 4.6.6.3 4.6.7.1 4.6.7.2 4.24.11.2 (Stud welding only)	_____ _____ _____ _____ _____ _____ _____ _____
Root treatment	4.6.9	_____
Other requirements	4.24.11.3 (Studs only) 4.24.11.4 (Studs only) 4.28 (Plug welds only) 4.29 (Slot welds only)	_____ _____ _____ _____

Specific requirements applicable to welding process (continued)

Submerged arc welding		
Procedure variable	Applicable provision	Indicate compliance
Filler metal: Specification	4.1.1, 4.1.4, 4.8.1	_____
classification	Tables 4.1.1, 4.1.4	_____
General requirements	4.7	_____
Fluxes	4.8, 4.9	_____
Single or multiple pass	4.1.5.2, 4.10, 4.11	_____
Single or multiple electrode	4.10, 4.11	_____
Welding current	4.10.3, 4.11.3	_____
Electrode restrictions	4.7.3	_____
	4.8.1	_____
	4.10.1	_____
	4.10.3	_____
	4.11.1	_____
	4.11.3.1(1)	_____
	4.11.3.1(2)	_____
	4.11.3.1(3)	_____
	4.11.3.1(4)	_____
	4.11.3.2	_____
	4.11.4	_____
	4.11.5	_____
Root pass and weld layer restrictions	4.10.1	_____
	4.10.2	_____
	4.11.1	_____
	4.11.2	_____

Specific requirements applicable to welding process (continued)

Gas metal arc and flux cored arc welding

Procedure variable	Applicable provision	Indicate compliance
Filler metal: Specification classification	4.1.1, 4.1.4 Tables 4.1.1, 4.1.4, 4.12	_____ _____
Electrode restrictions	4.12.1 4.12.2 4.12.3 4.14.1.1 4.14.1.2	_____ _____ _____ _____ _____
Root pass and weld layer restrictions	4.14.1.3 4.14.1.4	_____ _____
Shielding gas	4.13, 4.14.3	_____
Single or multiple pass	4.1.5.3, 4.1.5.4, 4.14.1.3, 4.14.1.4	_____
Welding current	4.14.1.5	_____
Root treatment	4.14.2	_____
General requirements	4.14.1.6 4.14.4	_____ _____
	4.28 (Plug welds only)	_____
	4.29 (Slot welds only)	_____

Prepared by _____

Manufacturer or contractor _____

Date _____

Part B
Sample Welding Forms

This appendix contains eight forms that the Structural Welding Committee has approved for the recording of procedure qualification, welder qualification, welding operator qualification, and tacker qualification data required by this code. Also included are laboratory report forms for recording the results of nondestructive examination of welds.

It is suggested that the qualification and NDT information required by this code be recorded on these forms or similar forms which have been prepared by the user. Variations of these forms to suit the user's needs are permissible. These forms are available from AWS.

PREQUALIFIED JOINT WELDING PROCEDURE
PROCEDURE SPECIFICATION

Material specification _____

Welding process _____

Manual or machine _____

Position of welding _____

Filler metal specification _____

Filler metal classification _____

Flux _____

Weld metal grade* _____

Shielding gas _____ Flow rate _____

Single or multiple pass _____

Single or multiple arc _____

Welding current _____

Polarity _____

Welding progression _____

Root treatment _____

Preheat and interpass temperature _____

Postheat treatment _____

*Applicable only when filler metal has no AWS classification.

WELDING PROCEDURE

Pass no.	Electrode size	Welding current		Travel speed	Joint detail
		Amperes	Volts		

This procedure may vary due to fabrication sequence, fit-up, pass size, etc., within the limitation of variables given in 4B, C, or D of AWS D1.1, Structural Welding Code.

Procedure no. _____ Manufacturer or contractor _____

Revision no. _____ Authorized by _____

Date _____

Form E-1

WELDING PROCEDURE QUALIFICATION TEST RECORD

PROCEDURE SPECIFICATION

Material specification _____
Welding process _____
Manual or machine _____
Position of welding _____
Filler metal specification _____
Filler metal classification _____
Weld metal grade* _____
Shielding gas _____ Flow rate _____
Single or multiple pass _____
Single or multiple arc _____
Welding current _____
Welding progression _____
Preheat temperature _____
Postheat treatment _____
Welder's name _____

*Applicable when filler metal has no
 AWS classification.

GROOVE WELD TEST RESULTS

Reduced-section tension tests

Tensile strength, psi

1. _____
2. _____

Guided-bend tests (2 root-, 2 face-, or 4 side-bend)

Root	Face
1. _____	1. _____
2. _____	2. _____

Radiographic-ultrasonic examination

FILLET WELD TEST RESULTS

Minimum size multiple pass Macroetch	Maximum size single pass Macroetch
1. ____ 3. ____	1. ____ 3. ____
2. ____	2. ____

All-weld-metal tension test

Tensile strength, psi _____
Yield point, psi _____
Elongation in 2 in., % _____

Laboratory test no. _____

WELDING PROCEDURE

Pass No.	Elect. size	Welding current		Speed of travel	Joint detail
		Amperes	Volts		

We, the undersigned, certify that the statements in this record are correct and that the test welds were prepared, welded, and tested in accordance with the requirements of 5B of AWS D1.1, Structural Welding Code.

Procedure no. _____   Manufacturer or contractor _____

Revision no. _____   Authorized by _____

Date _____

WELDING PROCEDURE QUALIFICATION TEST RECORD
FOR ELECTROSLAG AND ELECTROGAS WELDS

PROCEDURE SPECIFICATION

Material specification _____
Welding process _____
Position of welding _____
Filler metal specification _____
Filler metal classification _____
Filler metal _____
Flux _____
Shielding gas _____ Flow rate _____
Gas dew point _____
Thickness range this test qualifies _____
Single or multiple pass _____
Single or multiple arc _____
Welding current _____
Preheat temperature _____
Postheat treatment _____
Welder's name _____

TEST RESULTS

Reduced-section tension test

Tensile strength, psi
1. _____
2. _____

All-weld-metal tension test

Tensile strength, psi _____
Yield point, psi _____
Elongation in 2 in., % _____

Side-bend tests

1. _____ 3. _____
2. _____ 4. _____

Radiographic-ultrasonic examination _____

Impact tests

Size of specimen _____ Test temp. _____
Ft-lb: 1. _____ 2. _____ 3. _____ Avg. _____
High _____ Low _____

Laboratory test no. _____

WELDING PROCEDURE

Pass no.	Electrode size	Welding current		Joint detail
		Amperes	Volts	

Guide tube flux _____
Guide tube composition _____
Guide tube diameter _____
Vertical rise speed _____
Traverse length _____
Traverse speed _____
Dwell _____
Type of molding shoe _____

We, the undersigned, certify that the statements in this record are correct and that the test welds were prepared, welded, and tested in accordance with the requirements of 4E and 5B of AWS D1.1, Structural Welding Code.

Procedure no. _____ Manufacturer or contractor _____

Revision no. _____ Authorized by _____

Date _____

Form E-3

WELDER AND WELDING OPERATOR QUALIFICATION TEST RECORD

Welder or welding operator's name _____ Identification no. _____
Welding process _____ Manual _____ Semiautomatic _____ Machine _____
Position _____
(Flat, horizontal, overhead or vertical — if vertical, state whether upward or downward)
In accordance with procedure specification no. _____
Material specification _____
Diameter and wall thickness (if pipe) — otherwise, joint thickness _____
Thickness range this qualifies _____

FILLER METAL

Specification no. _____ Classification _____ F no. _____
Describe filler metal (if not covered by AWS specification) _____
_____
Is backing strip used? _____
Filler metal diameter and trade name _____ Flux for submerged arc or gas for gas metal arc or flux
_____ cored arc welding _____

Guided Bend Test Results

Type	Result	Type	Result

Test conducted by _____ Laboratory test no. _____
per _____

Fillet Test Results

Appearance _____ Fillet size _____
Fracture test root penetration _____ Macroetch _____
(Describe the location, nature, and size of any crack or tearing of the specimen.)
Test conducted by _____ Laboratory test no. _____
per _____

RADIOGRAPHIC TEST RESULTS

Film identification	Results	Remarks	Film identification	Results	Remarks

Test witnessed by _____ Test no. _____
per _____

We, the undersigned, certify that the statements in this record are correct and that the welds were prepared and tested in accordance with the requirements of 5C or D of AWS D1.1, Structural Welding Code.

Manufacturer or contractor _____

Authorized by _____

Date _____

Form E-4

TACKER QUALIFICATION TEST RECORD

Tacker's name _____ Identification no. _____

Welding process _____

Position _____
(Flat, horizontal, overhead, or vertical — if vertical, state whether upward or downward)

In accordance with procedure specification no. _____

Material specification _____

Diameter and wall thickness (if pipe) — otherwise, joint thickness _____

Filler Metal

Specification no. _____ Classification _____ F no. _____

Describe filler metal (if not covered by AWS specification) _____
_____

For Information Only

Filler metal diameter and trade name _____ Flux for submerged arc·or gas for gas metal arc
_____ or flux cored arc welding _____

Test Results

Appearance _____ Fillet size _____

Fracture test root penetration _____ Soundness _____
(Describe the location, nature, and size of any crack or tearing of the specimen.)

Test conducted by _____ Laboratory test no. _____
            per _____

We, the undersigned, certify that the statements in this record are correct and tested in accordance with the requirements of 5E of AWS D1.1, Structural Welding Code.

Manufacturer or contractor _____

Authorized by _____

Date _____

Form E-5

REPORT OF ULTRASONIC EXAMINATION OF WELDS

Project _____ **Report no.** _____

Weld identification _____
Material thickness _____
Weld joint AWS _____
Welding process _____
Quality requirements – section no. _____
Remarks _____

Line number	Indication number	Transducer angle	Leg*	Decibels				Defect					Discontinuity evaluation	Remarks
				Indication level	Reference level	Attenuation factor	Indication rating	Length	Angular distance (sound path)	Depth from "A" surface	Distance			
											From X	From Y		
				a	b	c	d							
1														
2														
3														
4														
5														
6														
7														
8														
9														
10														
11														
12														
13														

Notes:
1. In order to attain "Rating D"
   (A) With instruments with gain control, use the formula
       a - b - c = d
   (B) With instruments with attenuation control, use the formula
       b - a - c = d
   (C) "A + OR" sign must accompany the "D" figure unless "D" is equal to zero.
2. Distance from X is used in describing the location of a weld discontinuity in a direction perpendicular to the weld reference line. Unless this figure is zero, "A + OR" sign must accompany it.
3. Distance from Y is used in describing the location of a weld discontinuity in a direction parallel to the weld reference line. This figure is attained by measuring the distance from the "Y" end of the weld to the beginning of said discontinuity.
4. Make a separate report following repairs. (Suffix report no. with R1, R2, etc.)

*Use Leg I or II – see glossary of terms (Appendix I).

We, the undersigned, certify that the statements in this record are correct and that the welds were prepared and tested in accordance with the requirements of 6C of AWS D1.1, Structural Welding Code.

Inspected by _____ Manufacturer or contractor _____

Note: This form is applicable to Sections 8 and 9
(Buildings and Bridges). Do **NOT** use this form for   Authorized by _____
Tubular Structures (Section 10).

Date _____

Form E-6

REPORT OF RADIOGRAPHIC EXAMINATION OF WELDS

Project _____
Quality requirements – section no. _____
Reported to _____

WELD LOCATION AND IDENTIFICATION SKETCH

Technique
Source _____
Film to source _____
Exposure time _____
Screens _____
Film type _____

(Describe length, width, and thickness of all joints radiographed)

Date	Weld identification	Area	Interpretation		Repairs		Remarks
			Accept.	Reject	Accept.	Reject	

We, the undersigned, certify that the statements in this record are correct and that the welds were prepared and tested in accordance with the requirements of the American Welding Society Structural Welding Code, AWS D1.1.

Radiographer(s) _____   Manufacturer or contractor _____

Interpreter _____   Authorized by _____

Date _____

Form E-7

REPORT OF MAGNETIC PARTICLE EXAMINATION OF WELDS

Project _____

Quality requirements – section no. _____

Reported to _____

WELD LOCATION AND IDENTIFICATION SKETCH

Date	Weld identification	Area	Interpretation		Repairs		Remarks
			Accept.	Reject	Accept.	Reject	

We, the undersigned, certify that the statements in this record are correct and that the welds were prepared and tested in accordance with the requirements of the American Welding Society Structural Welding Code, AWS D1.1.

Inspector _____   Manufacturer or contractor _____

Date _____   Authorized by _____

Method of inspection   Date _____

☐ Dry   ☐ Wet   ☐ Residual   ☐ Continuous

Type of magnetizing current

☐ AC   ☐ DC   ☐ Half-wave

Form E-8

Appendix F
Weld Quality Requirements for Tension Joints in Bridges

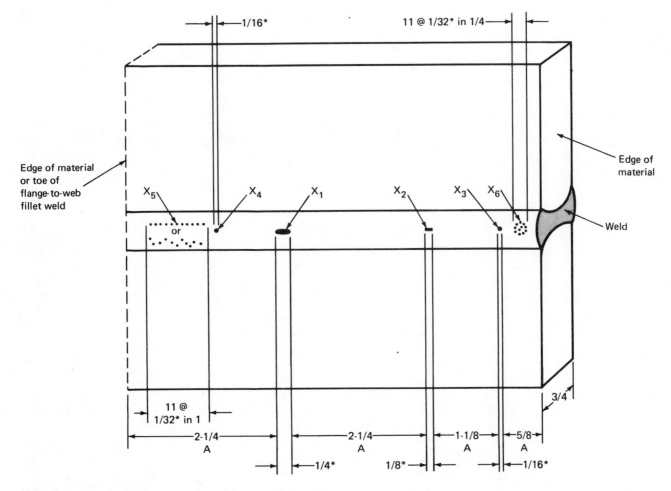

Notes:
1. A — minimum clearance allowed between edges of porosity or fusion-type discontinuities 1/16 in. (1.6 mm) or larger. Larger of adjacent discontinuities governs.
2. X_1 — largest permissible porosity or fusion-type discontinuity for 3/4 in. (19.0 mm) joint thickness (see Fig. 9.25.2.1).
3. X_2, X_3, X_4 — porosity or fusion-type discontinuity 1/16 in. (1.6 mm) or larger, but less than maximum permissible for 3/4 in. (19.0 mm) joint thickness.
4. X_5, X_6 — porosity or fusion-type discontinuity less than 1/16 in.

Interpretation:
1. Porosity or fusion-type discontinuity X_4 is not acceptable because it is within the minimum clearance allowed between edges of such discontinuities (see 9.25.2.1 and Fig. 9.25.2.1).
2. Remainder of weld is acceptable.

*Defect size indicated is assumed to be its greatest dimension.

Appendix G: Flatness of Girder Webs—Buildings

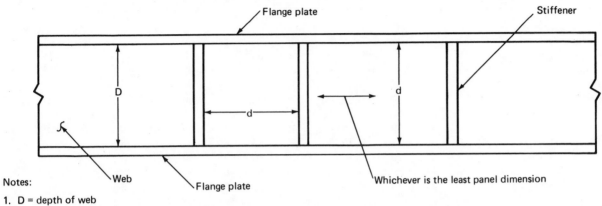

Flange plate

Stiffener

D

d

d

Whichever is the least panel dimension

Web

Flange plate

Notes:

1. D = depth of web
2. d = least panel dimension

Intermediate stiffeners on both sides of web, dynamic loading

Thickness of web	Depth of web	Least panel dimension, in.													
5/16	Less than 47	29	36	43	50										
	47 and over	23	29	35	40	46	52	58	63	69	75	81	86	92	98
3/8	Less than 56	29	36	43	50	58									
	56 and over	23	29	35	40	46	52	58	63	69	75	81	86	92	98
7/16	Less than 66	29	36	43	50	58	65								
	66 and over	23	29	35	40	46	52	58	63	69	75	81	86	92	98
1/2	Less than 75	29	36	43	50	58	65	72	79						
	75 and over	23	29	35	40	46	52	58	63	69	75	81	86	92	98
9/16	Less than 84	29	36	43	50	58	65	72	79	86					
	84 and over	23	29	35	40	46	52	58	63	69	75	81	86	92	98
5/8	Less than 94	29	36	43	50	58	65	72	79	86	93				
	94 and over	23	29	35	40	46	52	58	63	69	75	81	86	92	98
						Maximum permissible variation									
		1/4	5/16	3/8	7/16	1/2	9/16	5/8	11/16	3/4	13/16	7/8	15/16	1	1-1/16

Thickness of web, mm	Depth of web	Least panel dimension, meters													
8.0	Less than 1.19	0.74	0.91	1.09	1.27										
	1.19 and over	0.58	0.74	0.89	1.02	1.17	1.32	1.47	1.60	1.75	1.90	2.06	2.18	2.34	2.49
9.5	Less than 1.42	0.74	0.91	1.09	1.27	1.47									
	1.42 and over	0.58	0.74	0.89	1.02	1.17	1.32	1.47	1.60	1.75	1.90	2.06	2.18	2.34	2.49
11.1	Less than 1.68	0.74	0.91	1.09	1.27	1.47	1.65								
	1.68 and over	0.58	0.74	0.89	1.02	1.17	1.32	1.47	1.60	1.75	1.90	2.06	2.18	2.34	2.49
12.7	Less than 1.90	0.74	0.91	1.09	1.27	1.47	1.65	1.83	2.00						
	1.90 and over	0.58	0.74	0.89	1.02	1.17	1.32	1.47	1.60	1.75	190	2.06	2.18	2.34	2.49
14.3	Less than 2.13	0.74	0.91	1.09	1.27	1.47	1.65	1.83	2.00	2.18					
	2.13 and over	0.58	0.74	0.89	1.02	1.17	1.32	1.47	1.60	1.75	1.90	2.06	2.18	2.34	2.49
15.9	Less than 2.39	0.74	0.91	1.09	1.27	1.47	1.65	1.83	2.00	2.18	2.36				
	2.39 and over	0.58	0.74	0.89	1.02	1.17	1.32	1.47	1.60	1.75	1.90	2.06	2.18	2.34	2.49
						Maximum permissible variation, millimeters									
		6.4	8.0	9.5	11.1	12.7	14.3	15.9	17.5	19.0	20.6	22.2	23.8	25.4	27.0

Note: For actual dimensions not shown, use the next highest figure.

Intermediate stiffeners on both sides of web, static loading

Thickness of web	Depth of web	Least panel dimension, in.													
5/16	Less than 47	25	31	38	44	50									
	47 and over	20	25	30	35	40	45	50	55	60	65	70	75	80	85
3/8	Less than 56	25	31	38	44	50	56	63							
	56 and over	20	25	30	35	40	45	50	55	60	65	70	75	80	85
7/16	Less than 66	25	31	38	44	50	56	63	69						
	66 and over	20	25	30	35	40	45	50	55	60	65	70	75	80	85
1/2	Less than 75	25	31	38	44	50	56	63	69	75	81				
	75 and over	20	25	30	35	40	45	50	55	60	65	70	75	80	85
9/16	Less than 84	25	31	38	44	50	56	63	69	75	81	88			
	84 and over	20	25	30	35	40	45	50	55	60	65	70	75	80	85
5/8	Less than 94	25	31	38	44	50	56	63	69	75	81	88	94		
	94 and over	20	25	30	35	40	45	50	55	60	65	70	75	80	85
		Maximum permissible variation													
		1/4	5/16	3/8	7/16	1/2	9/16	5/8	11/16	3/4	13/16	7/8	15/16	1	1-1/16

Thickness of web, mm	Depth of web	Least panel dimension, meters													
8.0	Less than 1.19	0.63	0.79	0.97	1.12	1.27									
	1.19 and over	0.51	0.63	0.76	0.89	1.02	1.14	1.27	1.40	1.52	1.65	1.78	1.90	2.03	2.16
9.5	Less than 1.42	0.63	0.79	0.97	1.12	1.27	1.42	1.60							
	1.42 and over	0.51	0.63	0.76	0.89	1.02	1.14	1.27	1.40	1.52	1.65	1.78	1.90	2.03	2.16
11.1	Less than 1.68	0.63	0.79	0.97	1.12	1.27	1.42	1.60	1.75						
	1.68 and over	0.51	0.63	0.76	0.89	1.02	1.14	1.27	1.40	1.52	1.65	1.78	1.90	2.03	2.16
12.7	Less than 1.90	0.63	0.79	0.97	1.12	1.27	1.42	1.60	1.75	1.90	2.06				
	1.90 and over	0.51	0.63	0.76	0.89	1.02	1.14	1.27	1.40	1.52	1.65	1.78	1.90	2.03	2.16
14.3	Less than 2.13	0.63	0.79	0.97	1.12	1.27	1.42	1.60	1.75	1.90	2.06	2.24			
	2.13 and over	0.51	0.63	0.76	0.89	1.02	1.14	1.27	1.40	1.52	1.65	1.78	1.90	2.03	2.16
15.9	Less than 2.39	0.63	0.79	0.97	1.12	1.27	1.42	1.60	1.75	1.90	2.06	2.24	2.39		
	2.39 and over	0.51	0.63	0.76	0.89	1.02	1.14	1.27	1.40	1.52	1.65	1.78	1.90	2.03	2.16
		Maximum permissible variation, millimeters													
		6.4	8.0	9.5	11.1	12.7	14.3	15.9	17.5	19.0	20.6	22.2	23.8	25.4	27.0

Note: For actual dimensions not shown, use the next highest figure.

No intermediate stiffeners, dynamic or static loading

Thickness of web	Depth of web, in.																
Any	38	47	56	66	75	84	94	103	113	122	131	141	150	159	169	178	188
	Maximum permissible variation																
	1/4	5/16	3/8	7/16	1/2	9/16	5/8	11/16	3/4	13/16	7/8	15/16	1	1-1/16	1-1/8	1-3/16	1-1/4

Thickness of web, mm	Depth of web, meters																
Any	0.97	1.19	1.42	1.68	1.90	2.13	2.39	2.62	2.87	3.10	3.33	3.58	3.81	4.04	4.29	4.52	4.77
	Maximum permissible variation, millimeters																
	6.4	8.0	9.5	11.1	12.7	14.3	15.9	17.5	19.0	20.6	22.2	23.8	25.4	27.0	28.6	30.2	31.7

Note: For actual dimensions not shown, use the next highest figure.

Intermediate stiffeners on one side only of web, dynamic or static loading

Thickness of web	Depth of web	Least panel dimension, in.													
5/16	Less than 31	25	31												
	31 and over	17	21	25	29	34	38	42	46	50	54	59	63	67	71
3/8	Less than 38	25	31	38											
	38 and over	17	21	25	29	34	38	42	46	50	54	59	63	67	71
7/16	Less than 44	25	31	38	44										
	44 and over	17	21	25	29	34	38	42	46	50	54	59	63	67	71
1/2	Less than 50	25	31	38	44	50									
	50 and over	17	21	25	29	34	38	42	46	50	54	59	63	67	71
9/16	Less than 56	25	31	38	44	50	56								
	56 and over	17	21	25	29	34	38	42	46	50	54	59	63	67	71
5/8	Less than 63	25	31	38	44	50	56	63							
	63 and over	17	21	25	29	34	38	42	46	50	54	59	63	67	71
		Maximum permissible variation													
		1/4	5/16	3/8	7/16	1/2	9/16	5/8	11/16	3/4	13/16	7/8	15/16	1	1-1/16

Thickness of web, mm	Depth of web	Least panel dimension, meters													
8.0	Less than 0.78	0.63	0.79												
	0.79 and over	0.43	0.53	0.63	0.74	0.86	0.97	1.07	1.17	1.27	1.37	1.50	1.60	1.70	1.80
9.5	Less than 0.97	0.63	0.79	0.97											
	0.97 and over	0.43	0.53	0.63	0.74	0.86	0.97	1.07	1.17	1.27	1.37	1.50	1.60	1.70	1.80
11.1	Less than 1.12	0.63	0.79	0.97	1.12										
	1.12 and over	0.43	0.53	0.63	0.74	0.86	0.97	1.07	1.17	1.27	1.37	1.50	1.60	1.70	1.80
12.7	Less than 1.27	0.63	0.79	0.97	1.12	1.27									
	1.27 and over	0.43	0.53	0.63	0.74	0.86	0.97	1.07	1.17	1.27	1.37	1.50	1.60	1.70	1.80
14.3	Less than 1.42	0.63	0.79	0.97	1.12	1.27	1.42								
	1.42 and over	0.43	0.53	0.63	0.74	0.86	0.97	1.07	1.17	1.27	1.37	1.50	1.60	1.70	1.80
15.9	Less than 1.60	0.63	0.79	0.97	1.12	1.27	1.42	1.60							
	1.60 and over	0.43	0.53	0.63	0.74	0.86	0.97	1.07	1.17	1.27	1.37	1.50	1.60	1.70	1.80
		Maximum permissible variation, millimeters													
		6.4	8.0	9.5	11.1	12.7	14.3	15.9	17.5	19.0	20.6	22.2	23.8	25.4	27.0

Note: For actual dimensions not shown, use the next highest figure.

Appendix H: Flatness of Girder Webs—Bridges

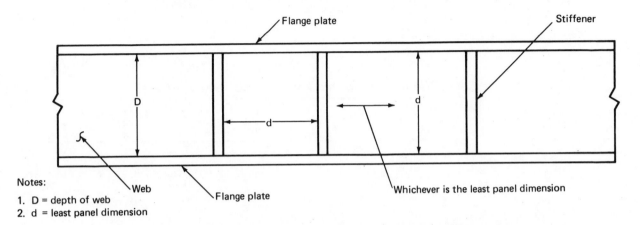

Notes:
1. D = depth of web
2. d = least panel dimension

Intermediate stiffeners on both sides of web, interior girders

Thickness of web	Depth of web	Least panel dimension, in.													
5/16	Less than 47	29	36	43	50										
	47 and over	23	29	35	40	46	52	58	63	69	75	81	86	92	98
3/8	Less than 56	29	36	43	50	58									
	56 and over	23	29	35	40	46	52	58	63	69	75	81	86	92	98
7/16	Less than 66	29	36	43	50	58	65								
	66 and over	23	29	35	40	46	52	58	63	69	75	81	86	92	98
1/2	Less than 75	29	36	43	50	58	65	72	79						
	75 and over	23	29	35	40	46	52	58	63	69	75	81	86	92	98
9/16	Less than 84	29	36	43	50	58	65	72	79	86					
	84 and over	23	29	35	40	46	52	58	63	69	75	81	86	92	98
5/8	Less than 94	29	36	43	50	58	65	72	79	86	93				
	94 and over	23	29	35	40	46	52	58	63	69	75	81	86	92	98
		Maximum permissible variation													
		1/4	5/16	3/8	7/16	1/2	9/16	5/8	11/16	3/4	13/16	7/8	15/16	1	1-1/16

Thickness of web, mm	Depth of web	Least panel dimensions, meters													
8.0	Less than 1.19	0.74	0.91	1.09	1.27										
	1.19 and over	0.58	0.74	0.89	1.02	1.17	1.32	1.47	1.60	1.75	1.90	2.06	2.18	2.34	2.49
9.5	Less than 1.42	0.74	0.91	1.09	1.27	1.47									
	1.42 and over	0.58	0.74	0.89	1.02	1.17	1.32	1.47	1.60	1.75	1.90	2.06	2.18	2.34	2.49
11.1	Less than 1.68	0.74	0.91	1.09	1.27	1.47	1.65								
	1.68 and over	0.58	0.74	0.89	1.02	1.17	1.32	1.47	1.60	1.75	1.90	2.06	2.18	2.34	2.49
12.7	Less than 1.90	0.74	0.91	1.09	1.27	1.47	1.65	1.83	2.00						
	1.90 and over	0.58	0.74	0.89	1.02	1.17	1.32	1.47	1.60	1.75	1.90	2.06	2.18	2.34	2.49
14.3	Less than 2.13	0.74	0.91	1.09	1.27	1.47	1.65	1.83	2.00	2.18					
	2.13 and over	0.58	0.74	0.89	1.02	1.17	1.32	1.47	1.60	1.75	1.90	2.06	2.18	2.34	2.49
15.9	Less than 2.39	0.74	0.91	1.09	1.27	1.47	1.65	1.83	2.00	2.18	2.36				
	2.39 and over	0.58	0.74	0.89	1.02	1.17	1.32	1.47	1.60	1.75	1.90	2.06	2.18	2.34	2.49
		Maximum permissible variation, millimeters													
		6.4	8.0	9.5	11.1	12.7	14.3	15.9	17.5	19.0	20.6	22.2	23.8	25.4	27.0

Note: For actual dimensions not shown, use the next highest figure.

Intermediate stiffeners on one side only of web, fascia girders

Thickness of web	Depth of web	Least panel dimension, in.													
5/16	Less than 31	30	38												
	31 and over	20	25	30	35	40	45	50	55	60	65	70	75	80	85
3/8	Less than 38	30	38												
	38 and over	20	25	30	35	40	45	50	55	60	65	70	75	80	85
7/16	Less than 44	30	38	45											
	44 and over	20	25	30	35	40	45	50	55	60	65	70	75	80	85
1/2	Less than 50	30	38	45	53										
	50 and over	20	25	30	35	40	45	50	55	60	65	70	75	80	85
9/16	Less than 56	30	38	45	53	60									
	56 and over	20	25	30	35	40	45	50	55	60	65	70	75	80	85
5/8	Less than 63	30	38	45	53	60	68								
	63 and over	20	25	30	35	40	45	50	55	60	65	70	75	80	85
		Maximum permissible variation													
		1/4	5/16	3/8	7/16	1/2	9/16	5/8	11/16	3/4	13/16	7/8	15/16	1	1-1/16

Thickness of web, mm	Depth of web	Least panel dimension, meters													
8.0	Less than 0.78	0.76	0.97												
	0.78 and over	0.51	0.63	0.76	0.89	1.02	1.14	1.27	1.40	1.52	1.65	1.78	1.90	2.03	2.16
9.5	Less than 0.97	0.76	0.97												
	0.97 and over	0.51	0.63	0.76	0.89	1.02	1.14	1.27	1.40	1.52	1.65	1.78	1.90	2.03	2.16
11.1	Less than 1.12	0.76	0.97	1.14											
	1.12 and over	0.51	0.63	0.76	0.89	1.02	1.14	1.27	1.40	1.52	1.65	1.78	1.90	2.03	2.16
12.7	Less than 1.27	0.76	0.97	1.14	1.35										
	1.27 and over	0.51	0.63	0.76	0.89	1.02	1.14	1.27	1.40	1.52	1.65	1.78	1.90	2.03	2.16
14.3	Less than 1.42	0.76	0.97	1.14	1.35	1.52									
	1.42 and over	0.51	0.63	0.76	0.89	1.02	1.14	1.27	1.40	1.52	1.65	1.78	1.90	2.03	2.16
15.9	Less than 1.60	0.76	0.97	1.14	1.35	1.52	1.73								
	1.60 and over	0.51	0.63	0.76	0.89	1.02	1.14	1.27	1.40	1.52	1.65	1.78	1.90	2.03	2.16
		Maximum permissible variation, millimeters													
		6.4	8.0	9.5	11.1	12.7	14.3	15.9	17.5	19.0	20.6	22.2	23.8	25.4	27.0

Note: For actual dimensions not shown, use the next highest figure.

Intermediate stiffeners on one side only of web, interior girders

Thickness of web	Depth of web	Least panel dimension, in.													
5/16	Less than 31	25	31												
	31 and over	17	21	25	29	34	38	42	46	50	54	59	63	67	71
3/8	Less than 38	25	31	38											
	38 and over	17	21	25	29	34	38	42	46	50	54	59	63	67	71
7/16	Less than 44	25	31	38	44										
	44 and over	17	21	25	29	34	38	42	46	50	54	59	63	67	71
1/2	Less than 50	25	31	38	44	50									
	50 and over	17	21	25	29	34	38	42	46	50	54	59	63	67	71
9/16	Less than 56	25	31	38	44	50	56								
	56 and over	17	21	25	29	34	38	42	46	50	54	59	63	67	71
5/8	Less than 63	25	31	38	44	50	56	63							
	63 and over	17	21	25	29	34	38	42	46	50	54	59	63	67	71
		Maximum permissible variation													
		1/4	5/16	3/8	7/16	1/2	9/16	5/8	11/16	3/4	13/16	7/8	15/16	1	1-1/16

Thickness of web, mm	Depth of web	Least panel dimension, meters													
8.0	Less than 0.78	0.63	0.79												
	0.79 and over	0.43	0.53	0.63	0.74	0.86	0.97	1.07	1.17	1.27	1.37	1.50	1.60	1.70	1.80
9.5	Less than 0.97	0.63	0.79	0.97											
	0.97 and over	0.43	0.53	0.63	0.74	0.86	0.97	1.07	1.17	1.27	1.37	1.50	1.60	1.70	1.80
11.1	Less than 1.12	0.63	0.79	0.97	1.12										
	1.12 and over	0.43	0.53	0.63	0.74	0.86	0.97	1.07	1.17	1.27	1.37	1.50	1.60	1.70	1.80
12.7	Less than 1.27	0.63	0.79	0.97	1.12	1.27									
	1.27 and over	0.43	0.53	0.63	0.74	0.86	0.97	1.07	1.17	1.27	1.37	1.50	1.60	1.70	1.80
14.3	Less than 1.42	0.63	0.79	0.97	1.12	1.27	1.42								
	1.42 and over	0.43	0.53	0.63	0.74	0.86	0.97	1.07	1.17	1.27	1.37	1.50	1.60	1.70	1.80
15.9	Less than 1.60	0.63	0.79	0.97	1.12	1.27	1.42	1.60							
	1.60 and over	0.43	0.53	0.63	0.74	0.86	0.97	1.07	1.17	1.27	1.37	1.50	1.60	1.70	1.80
		Maximum permissible variation, millimeters													
		6.4	8.0	9.5	11.1	12.4	14.3	15.9	17.5	19.0	20.6	22.2	23.8	25.4	27.0

Note: For actual dimensions not shown, use the next highest figure.

Intermediate stiffeners on both sides of web, fascia girders

Thickness of web	Depth of web	Least panel dimension, in.													
5/16	Less than 47	33	41	49											
	47 and over	26	33	39	46	53	59	66	72	79	85	92	98	105	112
3/8	Less than 56	33	41	49	57										
	56 and over	26	33	39	46	53	59	66	72	79	85	92	98	105	112
7/16	Less than 66	33	41	49	57	65	73								
	66 and over	26	33	39	47	53	59	66	72	79	85	92	98	105	112
1/2	Less than 75	33	41	49	57	65	73	81							
	75 and over	26	33	39	47	53	59	66	72	79	85	92	98	105	112
9/16	Less than 84	33	41	49	57	65	73	81	89						
	84 and over	26	33	39	47	53	59	66	72	79	85	92	98	105	112
5/8	Less than 94	33	41	49	57	65	73	81	89	98					
	94 and over	26	33	39	47	53	59	66	72	79	85	92	98	105	112
		Maximum permissible variation													
		1/4	5/16	3/8	7/16	1/2	9/16	5/8	11/16	3/4	13/16	7/8	15/16	1	1-1/16

Thickness of web, mm	Depth of web	Least panel dimension, meters													
8.0	Less than 1.19	0.84	1.04	1.24											
	1.19 and over	0.66	0.84	0.99	1.19	1.35	1.50	1.68	1.83	2.01	2.16	2.34	2.49	2.67	2.84
9.5	Less than 1.42	0.84	1.04	1.24	1.45										
	1.42 and over	0.66	0.84	0.99	1.19	1.35	1.50	1.68	1.83	2.01	2.16	2.34	2.49	2.67	2.84
11.1	Less than 1.68	0.84	1.04	1.24	1.45	1.65	1.85								
	1.68 and over	0.66	0.84	0.99	1.19	1.35	1.50	1.68	1.83	2.01	2.16	2.34	2.49	2.67	2.84
12.7	Less than 1.90	0.84	1.04	1.24	1.45	1.65	1.85	2.06							
	1.90 and over	0.66	0.84	0.99	1.19	1.35	1.50	1.68	1.83	2.01	2.16	2.34	2.49	2.67	2.84
14.3	Less than 2.13	0.84	1.04	1.24	1.45	1.65	1.85	2.06	2.26						
	2.13 and over	0.66	0.84	0.99	1.19	1.35	1.50	1.68	1.83	2.01	2.16	2.34	2.49	2.67	2.84
15.9	Less than 2.39	0.84	1.04	1.24	1.45	1.65	1.85	2.06	2.26	2.49					
	2.39 and over	0.66	0.84	0.99	1.19	1.35	1.50	1.68	1.83	2.01	2.16	2.34	2.49	2.67	2.84
		Maximum permissible variation, millimeters													
		6.4	8.0	9.5	11.1	12.7	14.3	15.9	17.5	19.0	20.6	22.2	23.8	25.4	27.0

Note: For actual dimensions not shown, use the next highest figure.

No intermediate stiffeners, interior or fascia girders

Thickness of web	Depth of web, in.																
Any	38	47	56	66	75	84	94	103	113	122	131	141	150	159	169	178	188
							Maximum permissible variation										
	1/4	5/16	3/8	7/16	1/2	9/16	5/8	11/16	3/4	13/16	7/8	15/16	1	1-1/16	1-1/8	1-3/16	1-1/4

Thickness of web, mm	Depth of web, meters																
Any	0.97	1.19	1.42	1.68	1.90	2.13	2.39	2.62	2.87	3.10	3.33	3.58	3.81	4.04	4.29	4.52	4.77
							Maximum permissible variation, millimeters										
	6.4	8.0	9.5	11.1	12.7	14.3	15.9	17.5	19.0	20.6	22.2	23.8	25.4	27.0	28.6	30.2	31.7

Note: For actual dimensions not shown, use the next highest figure.

Appendix I
Terms and Definitions

The terms and definitions in this glossary are divided into three categories: (1) general welding terms compiled by the AWS Committee on Definitions, Symbols, and Metric Practice; (2) terms, defined by the AWS Structural Welding Committee, which apply only to ultrasonic testing, designated by (UT) following the term; and (3) other terms, preceded by asterisks, which are defined as they relate to this code.

A

all-weld-metal test specimen. A test specimen with the reduced section composed wholly of weld metal.

amplitude length rejection level (UT). The maximum length of discontinuity permitted by various indication ratings associated with effective throat, as indicated in Tables 8.15.3 and 9.25.3.

angle of bevel. See preferred term **bevel angle.**

arc gouging. An arc cutting procedure used to form a bevel or groove.

as-welded. The condition of weld metal, welded joints, and weldments after welding prior to any subsequent thermal, mechanical, or chemical treatments.

attenuation (UT). The loss in acoustic energy which occurs between any two points of travel. This loss may be due to absorption, reflection, etc. (In this code, using the shear wave pulse echo method of testing, the attenuation factor is 2 dB per inch of sound path distance after the first inch.)

automatic welding. Welding with equipment which performs the welding operation without adjustment of the controls by a welding operator. The equipment may or may not perform the loading and unloading of the work. See **machine welding.**

axis of a weld. A line through the length of a weld, perpendicular to and at the geometric center of its cross section.

B

back gouging. The removal of weld metal and base metal from the other side of a partially welded joint to assure complete penetration upon subsequent welding from that side.

backing. Material (metal, weld metal, carbon, or granular) placed at the root of a weld joint for the purpose of supporting molten weld metal.

backing pass. A pass made to deposit a backing weld.

backing ring. Backing in the form of a ring, generally used in the welding of piping.

backing strap. See preferred term **backing strip.**

backing strip. Backing in the form of a strip.

backing weld. Backing in the form of a weld.

*backup weld (tubular structures). The initial closing pass in a complete joint penetration groove weld, made from one side only, which serves as a backing for subsequent welding but is not considered as a part of the theoretical weld (Fig. 10.13.1.1, Details C and D).

back weld. A weld deposited at the back of a single-groove weld.

base metal. The metal to be welded, soldered, or cut.

bevel angle. The angle formed between the prepared edge of a member and a plane perpendicular to the surface of the member.

boxing. The continuation of a fillet weld around a corner of a member as an extension of the principal weld.

*brace intersection angle, Θ (tubular structures). The acute angle formed between brace center lines.

butt joint. A joint between two members aligned approximately in the same plane.

butt weld. A weld in a butt joint.

C

*caulking. Plastic deformation of weld and base metal surfaces by mechanical means to seal or obscure discontinuities.

complete fusion. Fusion which has occurred over the entire base metal surfaces exposed for welding and between all layers and passes.

complete joint penetration. Joint penetration in which the weld metal completely fills the groove and is fused to the base metal throughout its total thickness.

*****complete joint penetration groove weld (buildings and bridges).** A groove weld which has been made from both sides or from one side on a backing having complete penetration and fusion of weld and base metal throughout the depth of the joint.

*****complete joint penetration groove weld (tubular structures).** A groove weld having complete penetration and fusion of weld and base metal throughout the depth of the joint or as detailed in Fig. 10.13.1.1. A complete penetration tubular groove weld made from one side only, without backing, is permitted where the size or configuration, or both, prevent access to the root side of the weld.

complete penetration. See preferred term **complete joint penetration.**

consumable guide electroslag welding. See **electroslag welding.**

continuous weld. A weld which extends continuously from one end of a joint to the other. Where the joint is essentially circular, it extends completely around the joint.

corner joint. A joint between two members located approximately at right angles to each other.

CO₂ welding. See preferred term **gas metal arc welding.**

crater. In arc welding, a depression at the termination of a weld bead or in the molten weld pool.

D

decibel (UT). The logarithmic expression of a ratio of two amplitudes or intensities of acoustic energy.

decibel rating (dB) (UT). See preferred term **indication rating.**

*****defect.** A discontinuity or discontinuities which by nature or accumulated effect render a part or product unable to meet minimum applicable acceptance standards or specifications. This term designates rejectability.

defect level (UT). See preferred term **indication level.**

defect rating (UT). See preferred term **indication rating.**

defective weld. A weld containing one or more defects.

depth of fusion. The distance that fusion extends into the base metal or previous pass from the surface melted during welding.

*****dihedral angle.** See **local dihedral angle.**

*****discontinuity.** An interruption of the typical structure of a weldment such as a lack of homogeneity in the mechanical or metallurgical or physical characteristics of the material or weldment. A discontinuity is not necessarily a defect.

downhand. See preferred term **flat position.**

E

*****edge angle, ω (tubular structures).** The acute angle between a bevel edge made in preparation for welding and a tangent to the member surface, measured locally in a plane perpendicular to the intersection line. All bevels open to outside of brace.

effective length of weld. The length of weld throughout which the correctly proportioned cross section exists. In a curved weld, it shall be measured along the axis of the weld.

*****electrogas welding.** A method of gas metal arc welding or flux cored arc welding in which molding shoes confine the molten weld metal for vertical position welding.

electroslag welding (ESW). A welding process wherein coalescence is produced by molten slag which melts the filler metal and the surfaces of the work to be welded. The weld pool is shielded by this slag which moves along the full cross section of the joint as welding progresses. The conductive slag is kept molten by its resistance to electric current passing between the electrode and the work.

consumable guide electroslag welding—A method of electroslag welding in which filler metal is supplied by an electrode and its guiding member.

F

faying surface. The mating surface of a member which is in contact or in close proximity with another member to which it is to be joined.

filler metal. The metal to be added in making a welded, brazed, or soldered joint. See **electrode, welding rod, backing filler metal, brazing filler metal, diffusion aid,** and **solder** in AWS A3.0.

flash. The material which is expelled or squeezed out of a weld joint and which forms around the weld.

flat position. The welding position used to weld from the upper side of the joint and the face of the weld is approximately horizontal.

flux cored arc welding (FCAW). An arc welding process which produces coalescence of metals by heating them with an arc between a continuous filler metal (consumable) electrode and the work. Shielding is provided by a flux contained within the electrode. Additional shielding may or may not be obtained from an externally supplied gas or gas mixture.

fusion. The melting together of filler metal and base metal, or the melting of base metal only, which results in coalescence. See **depth of fusion.**

***fusion-type discontinuity.** Signifies slag inclusion, incomplete fusion, incomplete joint penetration, and similar discontinuities associated with fusion.

fusion zone. The area of base metal melted as determined on the cross section of a weld.

G

gas metal arc welding (GMAW). An arc welding process which produces coalescence of metals by heating them with an arc between a continuous filler metal (consumable) electrode and the work. Shielding is obtained entirely from an externally supplied gas or gas mixture. Some methods of this process are called MIG or CO_2 welding (nonpreferred terms).

***gas pocket.** A cavity caused by entrapped gas.

gouging. The forming of a bevel or groove by material removal. See also **back gouging, arc gouging,** and **oxygen gouging.**

groove angle. The total included angle of the groove between parts to be joined by a groove weld.

***groove angle, Φ (tubular structures).** The angle between opposing faces of the groove to be filled with weld metal, determined after the joint is fitted up.

groove face. That surface of a member included in the groove.

groove weld. A weld made in the groove between two members to be joined.

H

heat-affected zone. That portion of the base metal which has not been melted, but whose mechanical properties or microstructure have been altered by the heat of welding, brazing, soldering, or cutting.

horizontal fixed position (pipe welding).
The position of a pipe joint in which the axis of the pipe is approximately horizontal, and the pipe is not rotated during welding. See Figs. 5.8.1A, 5.8.1B, and 5.8.1.2.

horizontal position.
  fillet weld. The position in which welding is performed on the upper side of an approximately horizontal surface and against an approximately vertical surface. See Figs. 5.8.1A, 5.8.1B, and 5.8.1.3.
  groove weld. The position of welding in which the axis of the weld lies in an approximately horizontal plane and the face of the weld lies in an approximately vertical plane. See Figs. 5.8.1A, 5.8.1B, and 5.8.1.1.

horizontal reference line (UT). A horizontal line near the center of the ultrasonic test instrument scope to which all echoes are adjusted for dB reading.

horizontal rolled position (pipe welding).
The position of a pipe joint in which the axis of the pipe is approximately horizontal, and welding is performed in the flat position by rotating the pipe. See Figs. 5.8.1A, 5.8.1B, and 5.8.1.2.

***hot-spot strain (tubular structures).** The cyclic total range of strain which would be measured at the point of highest stress concentration in a welded connection. When measuring hot-spot strain, the strain gage should be sufficiently small to avoid averaging high and low strains in the regions of steep gradients.

I

indication (UT). The signal displayed on the oscilloscope signifying the presence of a sound wave reflector in the part being tested.

indication level (UT). The calibrated gain or attenuation control reading obtained for a reference line height indication from a discontinuity.

indication rating (UT). The decibel reading in relation to the zero reference level after having been corrected for sound attenuation.

inert-gas metal-arc welding. See preferred term **gas metal arc welding.**

intermittent weld. A weld in which the continuity is broken by recurring unwelded spaces.

interpass temperature. In a multiple-pass weld, the temperature (minimum or maximum as specified) of the deposited weld before the next pass is started.

J

joint. The location of members or the edges of members which are to be joined or have been joined.

joint penetration. The minimum depth a groove or flange weld extends from its face into a joint, exclusive of reinforcement.

joint welding procedure. The materials and detailed methods and practices employed in the welding of a particular joint.

L

lap joint. A joint between two overlapping members.

layer. A stratum of weld metal consisting of one or more weld beads laid side by side.

leg (UT). The path the shear wave travels in a straight line before being reflected by the surface of material being tested. See sketch for leg identification. Note: Leg I plus leg II equals one V-path.

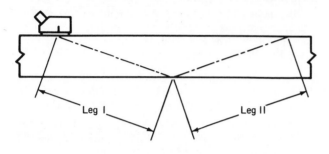

Leg I   Leg II

leg of a fillet weld. The distance from the root of the joint to the toe of the fillet weld.

*local dihedral angle, Ψ (tubular structures). The angle, measured in a plane perpendicular to the line of the weld, between tangents to the outside surfaces of the tubes being joined at the weld. The exterior dihedral angle, where one looks at a localized section of the connection, such that the intersecting surfaces may be treated as planes.

M

machine welding. Welding with equipment which performs the welding operation under the constant observation and control of an operator. The equipment may or may not perform the loading and unloading of the work. See **automatic welding.**

manual welding. A welding operation performed and controlled completely by hand. See **automatic welding, machine welding,** and **semiautomatic welding.**

N

node (UT). See preferred term **leg.**

O

overhead position. The position in which welding is performed from the underside of the joint. See Figs. 5.8.1A, 5.8.1B, 5.8.1.1, and 5.8.1.3.

overlap. The protrusion of weld metal beyond the toe, face, or root of the weld.

oxygen cutting (OC). A group of cutting processes used to sever or remove metals by means of the chemical reaction of oxygen with the base metal at elevated temperatures. In the case of oxidation-resistant metals, the reaction is facilitated by the use of a chemical flux or metal powder.

oxygen gouging. An application of oxygen cutting in which a bevel or groove is formed.

P

*parallel electrode. See submerged arc welding (SAW).

partial joint penetration. Joint penetration which is less than complete.

pass. A single longitudinal progression of a welding operation along a joint or weld deposit. The result of a pass is a weld bead or layer.

peening. The mechanical working of metals using impact blows.

plug weld. A circular weld made through a hole in one member of a lap or T-joint fusing that member to the other. The walls of the hole may or may not be parallel and the hole may be partially or completely filled with weld metal. (A fillet-welded hole or a spot weld should not be construed as conforming to this definition.)

*piping porosity (electroslag and electrogas). Elongated porosity whose major dimension lies in a direction approximately parallel to the weld axis.

*piping porosity (general). Elongated porosity whose major dimension lies in a direction approximately normal to the weld surface. Frequently referred to as "pin holes" when the porosity extends to the weld surface.

porosity. Cavity-type discontinuities formed by gas entrapment during solidification.

positioned weld. A weld made in a joint which has been so placed as to facilitate making the weld.

postweld heat treatment. Any heat treatment subsequent to welding.

preheating. The application of heat to the base metal immediately before welding, brazing, soldering, or cutting.

preheat temperature. A specified temperature that the base metal must attain in the welding, brazing, soldering, or cutting area immediately before these operations are performed.

procedure qualification. The demonstration that welds made by a specific procedure can meet prescribed standards.

Q

qualification. See preferred terms **welder qualification** and **procedure qualification.**

R

random sequence. See preferred term **wandering sequence.**

reference level (UT). The decibel reading obtained for

a horizontal reference line height indication from a reference reflector.

reference reflector (UT). The reflector of known geometry contained in the IIW Reference Block or other approved blocks.

reinforcement of weld. Weld metal in excess of the quantity required to fill a joint.

***rejectable discontinuity.** See preferred term **defect.**

resolution (UT). The ability of ultrasonic equipment to give separate indications from closely spaced reflectors.

root face. That portion of the groove face adjacent to the root of the joint.

root gap. See preferred term **root opening.**

root of joint. That portion of a joint to be welded where the members approach closest to each other. In cross section, the root of the joint may be either a point, a line, or an area.

root of weld. The points, as shown in cross section, at which the back of the weld intersects the base metal surfaces.

root opening. The separation, at the root of the joint, between the members to be joined.

S

scanning level (UT). The dB setting used during scanning, as described in Tables 8.15.3 and 9.25.3.

semiautomatic arc welding. Arc welding with equipment which controls only the filler metal feed. The advance of the welding is manually controlled.

shielded metal arc welding (SMAW). An arc welding process which produces coalescence of metals by heating with an arc between a covered metal electrode and the work. Shielding is obtained from decomposition of the electrode covering. Pressure is not used and filler metal is obtained from the electrode.

***shielding gas.** Protective gas used to prevent atmospheric contamination.

size of weld.
  groove weld. The joint penetration (depth of bevel plus the root penetration when specified).
  fillet weld. For equal leg fillet welds, the leg lengths of the largest isosceles right triangle which can be inscribed within the fillet weld cross section.

slot weld. A weld made in an elongated hole in one member of a lap or T-joint joining that member to that portion of the surface of the other member which is exposed through the hole. The hole may be open at one end and may be partially or completely filled with weld metal. (A fillet-welded slot should not be construed as conforming to this definition.)

sound beam distance (UT). See preferred term **sound path distance.**

sound path distance (UT). The distance between the search unit test material interface and the reflector as measured along the center line of the sound beam.

spatter. In arc and gas welding, the metal particles expelled during welding which do not form a part of the weld.

stringer bead. A type of weld bead made without appreciable weaving motion.

***stud base.** The stud tip at the welding end, including flux and container, and 1/8 in. (3.2 mm) of the body of the stud adjacent to the tip.

stud arc welding (SW). An arc welding process which produces coalescence of metals by heating with an arc between a metal stud or similar part and the other work part. When the surfaces to be joined are properly heated, heated, they are brought together under pressure. Partial shielding may be obtained by the use of a ceramic ferrule surrounding the stud. Shielding gas or flux may or may not be used.

submerged arc welding (SAW). An arc welding process which produces coalescence of metals by heating with an arc or arcs between a bare metal electrode or electrodes and the work. The arc is shielded by a blanket of granular, fusible material on the work. Pressure is not used and filler metal is obtained from the electrode and sometimes from a supplementary welding rod.
  ***single electrode.** One electrode connected exclusively to one power source which may consist of one or more power units.
  ***parallel electrode.** Two electrodes connected electrically in parallel and exclusively to the same power source. Both electrodes are usually fed by means of a single electrode feeder. Welding current, when specified, is the total for the two electrodes.

T

tack weld. A weld made to hold parts of a weldment in proper alignment until the final welds are made.

***tacker.** A fitter, or someone under the direction of a fitter, who tack welds parts of a weldment to hold them in proper alignment until the final welds are made.

***tandem.** Refers to a geometrical arrangement of electrodes in which a line through the arcs is parallel to the direction of welding.

temporary weld. A weld made to attach a piece or pieces to a weldment for temporary use in handling, shipping, or working on the weldment.

throat of a fillet weld.
  theoretical throat. The distance from the beginning of the root of the joint perpendicular to the hypotenuse of

the largest right triangle that can be inscribed within the fillet weld cross section.

actual throat. The shortest distance from the root of a fillet weld to its face.

throat of a groove weld. See preferred term **size of weld.**

T-joint. A joint between two members located approximately at right angles to each other in the form of a *T.*

toe of weld. The junction between the face of a weld and the base metal.

***transverse discontinuity.** A weld discontinuity whose major dimension is in a direction perpendicular to the weld axis "X", see Appendix E.

***tubular connection.** A connection in the portion of a tubular structure which contains two or more intersecting tubular members.

***tubular joint.** A joint in the interface created by one tubular member intersecting another.

U

undercut. A groove melted into the base metal adjacent to the toe or root of a weld and left unfilled by weld metal.

V

vertical position. The position of welding in which the axis of the weld is approximately vertical. See Figs. 5.8.1A, 5.8.1B, 5.8.1.1, and 5.8.1.3.

***vertical position pipe welding.** The position of a pipe joint wherein welding is performed in the horizontal position and the pipe shall not be rotated during welding. See Figs. 5.8.1A, 5.8.1B, and 5.8.1.2.

V-path (UT). The distance a shear wave sound beam travels from the search unit test material interface to the other face of the test material and back to the original surface.

W

wandering sequence. A longitudinal sequence in which the weld bead increments are deposited at random.

weave bead. A type of weld bead made with transverse oscillation.

weld. A localized coalescence of metals produced either by heating to suitable temperatures, with or without the application of pressure or by the application of pressure alone, and with or without the use of filler metal.

weldability. The capacity of a metal to be welded under the fabrication conditions imposed into a specific, suitably designed structure and to perform satisfactorily in the intended service.

weld bead. A weld deposit resulting from a pass. See **stringer bead** and **weave bead.**

welder. One who performs a manual or semiautomatic welding operation. (Sometimes erroneously used to denote a welding machine.)

welder certification. Certification in writing that a welder has produced welds meeting prescribed standards.

welder qualification. The demonstration of a welder's ability to produce welds meeting prescribed standards.

welding. A metal joining process used in making welds. (See the Master Chart of Welding Processes, AWS A3.1).

welding machine. Equipment used to perform the welding operation. For example, spot welding machine, arc welding machine, seam welding machine, etc.

welding operator. One who operates machine or automatic welding equipment.

welding procedure. The detailed methods and practices including all joint welding procedures involved in the production of a weldment. See **joint welding procedure.**

welding sequence. The order of making the welds in a weldment.

weldment. An assembly whose component parts are joined by welding.

Appendix J: Temperature–Moisture Content Charts

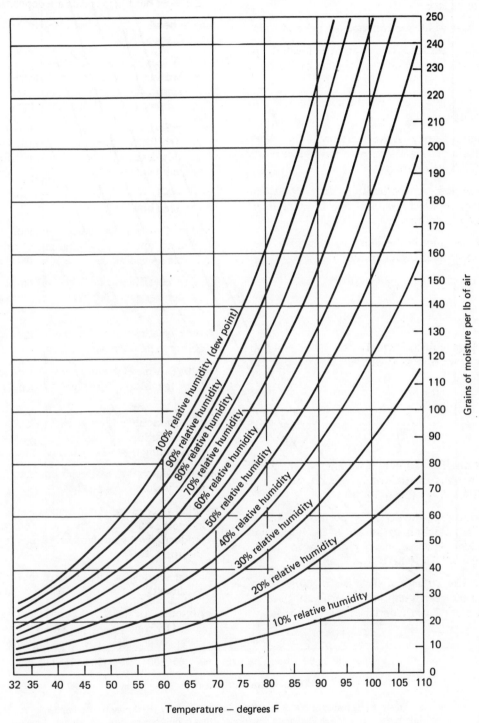

Notes:
1. Any standard psychrometric chart may be used in lieu of this chart.
2. See Fig. J2 for an example of the application of this chart in establishing electrode exposure conditions.

Fig. J1—Temperature-moisture content chart to be used in conjunction with testing program to determine extended atmospheric exposure time of low hydrogen electrodes (see 4.5.2)

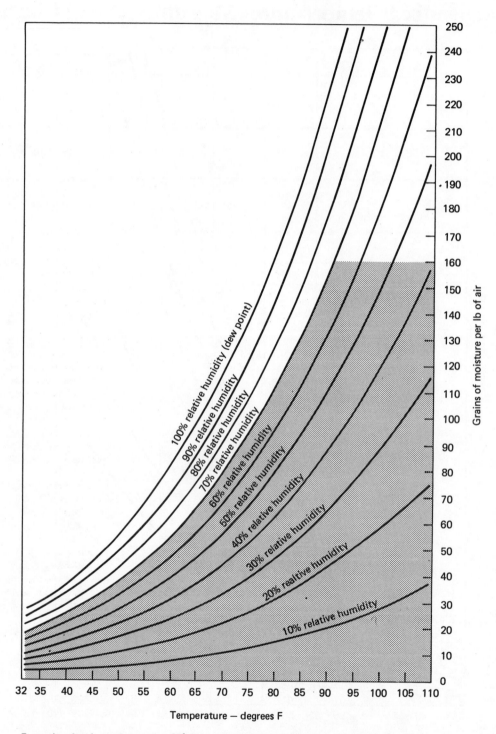

Temperature — degrees F

Example: An electrode tested at 90° F and 70% relative humidity (RH) may be used under the conditions shown by the shaded areas. Use under other conditions requires additional testing.

Fig. J2—Application of temperature-moisture content chart in determining atmospheric exposure time of low hydrogen electrodes

INDEX